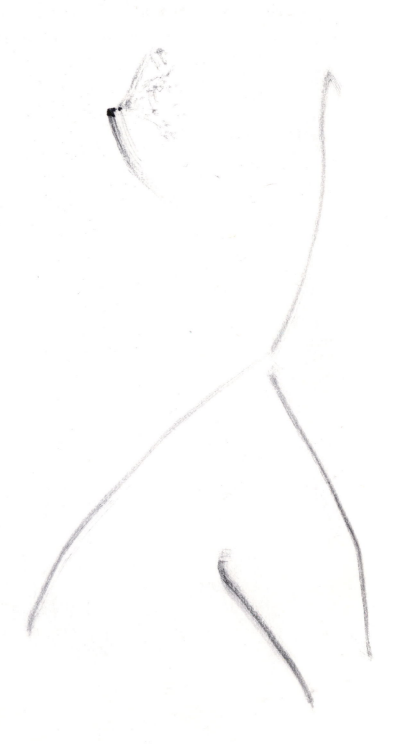

Concert Lighting

Concert Lighting

Techniques, Art, and Business

James L. Moody

Focal Press
Boston London

Focal Press is an imprint of Butterworth-Heinemann.

Library of Congress Cataloging-in-Publication Data

Moody, James L.
    Concert lighting : techniques, art, and
business / by James L. Moody.
        p.   cm.
    Bibliography: p.
    Includes index.
    ISBN 0–240–80010–9
    1. Music-halls—Lighting. 2. Lighting—
Special effects. 3. Stage lighting. I. Title.
TK4399.T6M66 1989
725′.81—dc19   88–16351
                                            CIP

British Library Cataloguing in Publication Data

Moody, James L.
    Concert lighting: techniques, art, and business
    1. Concerts, Stage lighting
    I. Title
    780′.7′3

    ISBN 0–240–80010–9

Butterworth-Heinemann
313 Washington Street
Newton, MA 02158

10 9 8 7 6 5 4 3 2
Printed in the United States of America

Dedicated to

all the people who have worked with
Sundance over the years and helped me
learn and enjoy a wonderfully exciting craft;
also, to the people who assisted with this
book: Ken Lammers, Scott Stipetic, Jeff
Ravitz, Chip Mounk, Chip Largman, Leo
Bonomy, and Dan Wohleen; finally, to the
person who believed I had something to
say and encouraged me to write, Joe Tawil.

Contents

Contents

Contents

Contents

Foreword

Jim Moody is the ideal person to write about concert business and lighting. Not only is he one of the entrepreneurs in this new lighting industry, but he writes with great clarity, organization, and readable simplicity.

His specialized knowledge of lighting for the concert field embraces both the technical and the artistic, as well as pragmatic organizational know-how. With a solid background in theatre lighting, degrees from Southern Illinois University and UCLA, he moved into concert lighting. He has been a leading advocate for fusion of the innovations and techniques of concert lighting into other production forms.

He was honored in 1980 with the first Lighting Designer of the Year Award presented by *Performance* magazine, the leading publication on concert touring. The award was based on a readers' poll of concert industry contemporaries. Jim has also received awards in television as well as theatre.

Those who are specialists in other lighting design areas—drama, musicals, opera, dance, television, amusement parks, landscape, and architectural lighting, and educators in these fields—will profit from reading this volume carefully. These days almost all working lighting designers cross over into all areas of design use in light.

Jim Moody is in a unique position to contribute to our knowledge by sharing his experience and insight into concert lighting. We are fortunate that he has done so in this volume.

Lee Watson
Purdue University
West Lafayette, Indiana

Preface

The Full Circle

Now that concert lighting has had over twenty years of growth, it is finally coming to the attention of a wide group of theatrical practitioners. I use the word *theatrical* to encompass theatre, dance, television, film, audiovisual, and corporate, as well as the fringe design media such as architectural and display lighting.

These experimental years have produced many useful techniques for many of the other theatrical lighting media. Not only were the developers of concert lighting interested in new technology, but also in the adaptation of older techniques from other media. They recognized that the past is often a greater teacher than the future and can be used to solve problems. This is well documented with what the early concert designers did to bring about this new art form. Theatre has traditionally borrowed from other fields. In its early days, television used many standards already set by the film industry. Now, all of these disciplines can gain from the innovations developed in this newest design field.

While introducing the reader to concert lighting, I will also show techniques that have application in the legitimate theatre and dance, as well as in television and film. However, I can only present the tip of the iceberg. Your individual needs and facilities will very likely require expansion or even totally new approaches to be tailored to your needs. I can only hope that this book will open the door to the experimentation that comes from the concert lighting experiences I have had as they are outlined in the following pages. You will also find examples of how they have already been used in some of these other media.

In creating a new theatrical form there had to be a base, a starting place. Much of it came from traditional theatre lighting. Theatre had the general form and style desired by the beginning concert designer, and it was the greatest exponent in the use of color media, a major concern for rock and roll designers. One great contribution, the PAR-64 fixture, came from film. Television was still heavily relying on film instrumentation when concert lighting began, so it contributed very little. But, as you will see, live video has become a big part of concerts and therefore the need to understand video is a must for the concert designer.

Concert lighting grabbed onto the principle of psychovisual relationships, the emotional impact of how people react to colored light, and expanded our use of its ability to influence the audience. The use of color became an art form unto itself. Whereas pastels and lightly tinted colors (no-colors) were the generally accepted rule in theatre, concert designers went for the most heavily saturated colors: bold, eye-catching, primary colors. The theatre textbooks on color do give the lighting designer the basic form from which to start, and should be studied before attempting to read this book. After all, we could not change the laws of physics; we just gave the audience new visual priorities.

The Broadway musical comes closest to what the concert lighting designer was looking for within the live theatre form. The followspot as a key source of light was relied on as the primary front light. But, ratio and relationship of side and back light took on new importance. Back light with the heavily saturated primary colors became the dominant source for mood and area lighting.

The available theatrical lighting fixtures in the mid-1960s had created problems for the concert lighting designer. The glass lenses did not troupe well, they were heavy, broke easily, and their lumen-per-watt ratio was poor. That term is used to give the designer or electrician a scale for determining how efficient the fixture is in relationship to other fixtures the designer has to choose from for the job at hand. Power availability was a major concern in the "found space" that concerts were regulated to perform within. They usually lacked the facilities for a normal theatre or studio setup. A more efficient light source was critical if the limited power resource was to be used.

The development of portable lighting structures will ultimately be the greatest single contribution concerts make to other lighting fields. This was a field virtually undeveloped by theatre, film, or television. The highly portable structures seen in concert work have become the basis for design research on new television studio lighting at several major facilities. The reduction in labor required to assemble and make operational these large groupings of lamps cannot be ignored in today's economy. The basic research and the hundreds of thousands of hours of on-the-job trial and error must be taken advantage of by the facility planners of the future.

The truss developed by concert people should not be viewed as a classic engineering model, but as a practical, everyday, structural concept that has now been time tested. The size and weight bearing factors will change with application, but the surprisingly utilitarian construction of the devices must be studied and refined if we are to continue to develop new staging techniques no matter what the field of use.

In the final analysis, what we do in any theatrical form is attempt to use light to influence the viewers' reactions to the scene, whether we are making a highly dramatic statement or simply providing the scene with enough illumination for the viewer to interpret the action. I still go by a saying I heard when I was in college:

> It's not where you put the light, it's where you don't put the light that makes for good theatrical lighting.

That line has always been a guidepost for me in my design for any project, no matter what the media. The absence of light is largely perceived as the dramatic key to the theatricality of creative lighting. It is what separates it from the normal illumination we find in our everyday lives.

It is incorrect to think that television lighting is nothing more than a technical function of the camera. The need for lighting to create mood and texture is very important. Beyond these, the need to create depth of field and contrast must always be considered. Chapter 19 will explain in detail the parameters for video and film reproduction. As with many points in the rest of the book, it is worth keeping these needs in mind for any theatrical lighting design.

The sum total of our experiences contributes to how we use lighting equipment in our various design applications. With the new experiences of concert lighting to draw from, the answer to a problem may be that much closer. A solution may be suggested in the following pages that will be of help sometime in the future.

This book presents a new lighting concept other than stage lighting. It is an introduction for students and others wishing to learn about concert lighting as a potential vocation, as well as a guide for designers and technicians in other fields. Those already involved with concert production will also find this a useful source. Understanding how concert lighting techniques were developed and how they can be applied in other media can expand any design's horizons. If we look outside our own limited experiences to the loosely connected media that also use a theatrical approach to lighting, we can enrich our creativity.

Concert lighting has often been thought of as unconventional, but a definition of unconventional use of lighting should not be limited to the use of "nontheatrical fixtures" such as neon, arc, or fluorescent sources. All theatrical lighting elements have unconventional applications if we think about a problem only in the broader context. Do not be limited by how someone tells you to use something or how it was used in the past. Take the situation at hand and attempt to solve the problem using whatever means it takes. Theatrical lighting is a creative technology, not an exacting science.

In the following chapters I use the experiences I have had during my seventeen years in the concert lighting field. This book is not meant to be the definitive work on the subject, but a highly personal view of a few of my designs and my involvement in concert lighting as well as other media. How I have mixed them all together to develop unconventional approaches and techniques in lighting is a big part of my technique. This practical, hands-on approach has been uppermost in my mind while I have been writing and for this reason some reiteration may be evident from time to time. This is intentional, for it is sometimes tedious to have to refer back to previous pages. Furthermore, if the basic ideas are restated in several places, it will help you to remember them and their paramount importance becomes more evident.

Some new terms may not be followed by definitions (you can check the Glossary). I am assuming that the reader has already obtained a basic knowledge of lighting before reading this volume.

Accept my ideas in the spirit in which they were written, simply as a new way to see things. Use the good ideas found in life and discard the bad has always been my philosophy. Make it yours, turn it around, and look at it in new and different ways. Make your own analysis, then push forward and develop new innovations of your own.

I have also included several chapters which on first glance may seem unrelated to the specific job of lighting a concert. It takes more than a knowledge of lumens and color charts to be a professional designer. Ours is a collaborative art and the inner workings of the whole production as it relates to your specific job is essential. Since the concert media has not been written about extensively, I felt I had to spend a portion of this book putting the field into perspective for the reader. Read these chapters as if they were the key to the outer door. You cannot get into the inner sanctum without it.

Acknowledgments

I am indebted to the past and present publishers of *Lighting Dimensions*, Fred Weller and Patricia McKay, editor Arnold Aronson of *Theatre Design and Technology*, and Patricia McKay as editor and publisher of *Theatre Crafts* for their courteous permission to use material that appeared in articles I contributed to those periodicals.

James L. Moody
Los Angeles, California
January 1989

I

The Concert Lighting Field

1

The Birth of Rock and the Rise of the Concert Lighting Designer

It is hard to pinpoint the actual beginning of concert lighting as we think of it today. Certainly the *Grand Tour* could be seen as having been the byproduct of opera in the mid-1800s. The term was often given to a star's travels through Europe, presenting solo programs in the major European cultural capitals. Later, the Grand Tour came to the Americas. Through the years it also came to include the popular figures of show business, encompassing not only opera but the stars of dance hall, vaudeville, and the circus. In the late 1800s, despite their isolated locations, even small Nevada gold rush towns had an Opera House to show the world how "cultured" they had become.

The swing bands of the 1920s and 1930s brought a big change to popular music and, some believe, sounded the first notes that would ultimately be recognized as *Rock and Roll*. Led by such greats as Duke Ellington, Count Basie, and Paul Whiteman, they emphasized instrumental solos—riffing, or playing a short phrase over and over, now considered a key ingredient in rock and roll.

Another milestone was the entrance of the *pop idol*. Although Benny Goodman is widely credited for igniting the first "teen hysteria" in 1938 at a Carnegie Hall concert, it would later be a teenager from Hoboken, New Jersey—Francis Albert Sinatra—who was to sustain a legion of young teenage girls screaming during his performances.

The melding of country music and such regional sounds as rhythm and blues had been building until in 1951 a disc jockey named Alan Freed started the "Moondog Show" on WJW radio in Cleveland, Ohio. In 1954 the name was changed to "The Rock and Roll Show." *Rolling Stone* magazine wrote that the term was perfect because:

> It was a way of distinguishing the new rhythm and blues from just plain blues and the old corny Mills Brothers style. After all, rock and roll didn't fit into any of the old categories. . . . (*Rock of Ages, the Rolling Stone History of Rock & Roll*, Rolling Stone Press, New York, 1986, p. 96–97).

The benchmark of modern concert touring was set in the mid-1950s by the independent record companies in an effort to exploit the fledgling rock and roll recording artists. They were not a very radical departure from what swing bands and orchestras had been doing in the 1930s, 1940s and 1950s; that is, playing dances in every town that had a community hall or theatre. After all, this was the way most musicians made their living—playing live dances. But now the pop singer moved from being just a member of the band to fronting it instead of the bandleader. Another change was that instead of the band getting work through a booking agent who was separate from their manager, the pop singers were promoted by an independent record producer who also controlled the record, or sometimes an independent entrepreneur like Alan Freed. That way the record company not only made money from the ticket sales, but, more important, by stimulating record sales. Many artists signed away their publishing rights for a small, onetime fee or were lied to by the record companies concerning the record's earnings.

Concert Lighting Begins

Lighting did not come into a prominent position until after sound reinforcement made the first inroads about 1960. The inadequate sound system in most buildings could not handle the demands of the recording artists who had come to expect studio quality sound (not to mention the new electronic effects necessary to make them sound like the record). After the artist got used to absorbing the expense of carrying sound equipment from city to city, lighting was soon to follow.

One of the first artists to do this was Harry Belafonte in the mid-1960s. He had come on the record scene in 1957 from his native home in Jamaica and was truly ahead of his time. Generally, the MOR (Middle-Of-the-Road or light rock) and country/western artists were the last to see the value of building a production around their music.

The Light Show

What became known as 1960s acid rock was spearheaded by such bands as Big Brother and the Holding Company, Jefferson Airplane, Warlock, and Quicksilver Messenger Service. All were San Francisco–based. Actually it was a nonmusical group, The Family Dog, at the close of 1965, that created the first *light show*. Light shows were put on as part of events called "Happenings," which included films, dance, music, mime, painting, and just about anything else people wanted to do, all going on at once.

Bill Graham had just started managing the San Francisco Mime Troupe and was brought in to produce the now famous Trips Festival at the Longshoremen's Hall in January of 1966. Later that year, believing that music could be promoted without the other elements, he rented the Fillmore Theatre on his own and started promoting concerts that featured individual bands as the main attraction without the ancillary features. Films, however, were shown during the set changes to keep the audience occupied.

At first the light shows were staged by the promoter, not the bands. They were a visual explosion of color and design. The show was based on liquid light projections, strobe lights, black light, and effect lighting to create a visual mood as an environment into which the band as well as audience was immersed. The liquid light projector was nothing fancier than the opaque projector your grade school teacher used to show photos and charts from books. Only now in place of the book was a pan with oil or water into which paints were pushed, splashed, and injected. The pan was vibrated or tilted to add even more movement to the ever-changing patterns that this mixture created.

Melding Forms

The concerts moved from light shows to more traditional music hall lighting simply because certain performers began to emerge from these groups and gain star status. With this came audience recognition. I believe that the artist's ego was responsible for a move toward the more standard musical comedy lighting techniques. They wanted to be in the spotlight, the traditional symbol of the "star."

Like any melding of the old with the new, and especially in light of the 1960s' youth revolt, these record artists started altering the rules. There wasn't a comedian or animal act to separate the musical presentations as was common in music halls or vaudeville. This was to be a whole show of solid music, lasting several hours. Lighting began to take on a more important role.

The greatest differences in concert lighting from traditional theatrical lighting are: the use of vivid colors; heavy use of back light; and absolute use of followspots instead of balcony rail, torms, or front-of-house washes. The greatest advances have been in portable lighting structures, the PAR-64 fixture, and computerized moving lights.

The Concert Lighting Designer/Director

The position of a concert lighting designer/director differs from its theatrical or television counterpart. This mainly centers around responsibilities. The concert lighting designer is often the only design artist associated with the production. Only the larger concerts can afford a separate scenic artist, so the lighting designer is usually consulted for all visual concepts. Second, there is rarely an extended rehearsal schedule. Often a lighting designer will have only one day with the lighting rig. No thought of a stop-and-go technical rehearsal here.

Concert lighting designers must have a highly developed musical sense. Although many are not skilled musicians, they have a natural aptitude for musical interpretation. Because of the ever-changing venues and artist needs, most of them go on the road with the shows they design. A few do leave the show with a board operator, but I find most designers stay and personally run the console and call their own cues throughout the tour.

This is an immediacy art! You are not an artist who can put paint on canvas or chip a piece of stone and then stand back to think about your next move for an hour, a day, or a week—you must react instantly. Often there is no time to write down what you did.

Concert lighting design is an intuitive art, never to be exactly reproduced again. There are no cue sheets or scripts to make notations on here. Ways of noting lighting cues have been developed but in a much different form than the standard theatrical method. Chapter 6 will detail one such method.

Preparation is the key. Every day brings new locations and, therefore, a new set of problems to solve.

Adaptability is a must. One of the most important lessons to learn is:

> There is no such thing as a bad decision; the only wrong decision is to make no decision.

If you are prepared for all conceivable problems, then you can deal logically and calmly with the everyday disasters. Although a lot of innovations have been tried in the last twenty-odd years of concert lighting, the techniques are still evolving. Just when we feel the size of the lighting rigs has been pushed to the limit, a new idea is tried that pushes the physical resources of the media and particularly the physical structure of the buildings, as well as our imaginations.

I do not believe in an analytical approach—in this business it takes ¼ art, ¼ science, ¼ intuition, and ¼ adaptability. Teaching by doing, experimenting, and by seeing what others have tried is of the greatest value. That is why this book is presented in four main parts. Part I deals with what you should understand about the work: the business and the physical side as well as the creativity it takes to succeed. I have a very strong conviction that designers need to be well versed in business to get ahead in the real world. Part II shows some of the tools presently being used. Part III is a look at some of my own designs. They reflect what I did to solve both the creative and business needs of the specific tour. They are, in my view, inseparable. Part IV shows how I have used these techniques and other ideas gained in my years of touring to cross over into other media. Finally, Part V, the Afterword, presents opinions from four other top designers in concert lighting and an overview of overseas touring.

2

Tour Personnel and Unions

Anyone following the business of rock and roll touring has probably become familiar with the term *roadie*. I would like to counter the stereotype that this term has established, and discourage its use in the future. The term roadie itself carries a certain degrading connotation, having its roots in the tradition that spawned another infamous rock and roll term—*groupie*. Many road crew members in the early years may have been little more than male groupies (family and friends of band members) who simply wanted to hang around with the band, but the industry and its requirements have changed a lot in the past decade. Rock and roll as a whole has become more sophisticated and technically more complex, and as a result, the persons charged with the care and handling of equipment have become highly important to the success of a tour. Road crew members have evolved from hangers-on into trained technicians with specific expertise in electronics, musical instrument repair, lighting, sound, and the allied theatrical arts. The untrained hanger-on of the early years has been replaced, for the most part, by the dedicated, trained, touring professional of today, whom I prefer to call the *equipment manager* or *technician*.

There are few standards or formulas in this business. Crew size, wages, titles, responsibilities, and equipment complexity are variable factors that depend on the nature of the show, the whims of the artists and their management, and the financial limits of the tour. The titles and duties listed here are general definitions as applied to rock and roll touring and are often subject to adaptation.

Road Crew Duties

Generally, a touring artist has a manager, a road manager, one or more truck drivers, a band, and lighting and sound technicians. As an artist's earning power increases, the productions often become more and more elaborate and the technical staffs increase in size. So, in addition to the basic staff listed above, a touring act may also have: security, PR (public relations), stage manager, pyrotechnician, rigger, audiovisual specialist, laserist, set designer, staging company, tour accountant, costumer, carpenter, or any of the other standard theatrical titles.

A look at Figure 2-1 will confirm how many positions may need to be filled to get a tour on the road.

Manager

Like their namesakes in the theatre, managers are closest to the performers, often handling their contracts, bookings, and money as well as acting as the performer's confidant in personal matters. Managers seem to come in an endless variety of styles: some are deeply concerned with the physical production, some are only happy if the artist

Figure 2-1 Backstage
How many people it takes to mount a tour is well illustrated in this example from the September 4th, 1987 issue of *Performance* magazine's "Backstage." (Reprinted from vol. 17, no. 16, p. 9, by permission of the publisher.)

is happy, others don't want to spend any money, but still demand a topnotch production. Luckily, this combination IBM 360 computer and surrogate mother is usually too busy in the office or managing other artists to tour with the show.

Road Manager

The road manager is in charge of the tour. The primary responsibilities of the job are to keep the artists happy, functional, and performing; to see that the show goes on no matter what; and, on all but the big tours, to settle the box office and carry cash for the expenses. Many road managers have hopes of becoming managers someday, having worked their way up from crew positions. He or she will be heavily involved in the pretour planning of schedules and tour arrangements and probably has final approval in hiring road crew members.

Tour Accountant

A relatively new face on tours in the 1980s is a person trained in accounting. He or she not only collects the box office and pays out per diems, but settles the really big money—T-shirts and poster sales.

Seriously, the monies that are involved in an arena-type show and the very complicated splits of gross almost dictate that a specialist is needed full-time on the road just to deal with monies.

Security

As the title implies, security personnel protect the artist, and keep unknown or unwanted persons off the stage and out of the backstage areas. Most bodyguards are very personable, and act more like valets than anything else; few carry guns, but most are experts in the martial arts. Frankly, everyone feels a little safer with these people around. It is an unfortunate fact that there are a few persons out there that do want to harm performers. There is a tremendous investment in the artist and a lot of people are counting on the artist's ability to perform to secure their own livelihoods.

Public Relations

Often the PR or press agent works for an outside firm hired by the artist or record company to handle publicity. PR people are most concerned with "image and publicity." The need for any performer to keep himself in the public eye makes it essential that a coordinated effort be made to ensure not only that the local press has the artist's name on their lips, but that it is a positive image so that "good press" is the result.

Truck Driver

Many drivers and trucking companies specialize in hauling for concert tours. The drivers live by one rule: Get the equipment to the next hall no matter what! They are usually young and nonunion, carrying the necessary Class-A driver's license for tractor-trailer rigs, and may or may not help with the load-out each night. There are major trucking companies with union drivers who do a great job, but the percentages are against them. Good truck drivers are highly valued because if they are late to the hall many crew hours are wasted. In the worst case, a delayed truck could cause the show to be cancelled. Usually drivers are part of the transportation package supplied by one of the many companies specializing in touring, but some bands who tour consistently keep the same drivers on a retainer. Since a retainer is rare for the sound and lighting crew, it shows just how valuable truck drivers are to a tour.

Lighting Designer

The touring LD (lighting designer) strives to meet all of the usual demands of creative lighting, but with an added twist: unlike theatre, which offers a script and a directorial concept for guidance, rock acts rarely have a concept, a program, or a director. At best, the director is the artist, whose onstage point of view offers little in the way of objective criticism. The LD often runs the control console personally and may be involved in the physical work of erecting the lighting rig during the setup. Rehearsal time before leaving town is limited, and more often nonexistent, so the LD must be able to improvise at the first few shows. And while trying to second-guess the act as to what song will be played next, and preset the two-scene or memory console, the LD must also cue the followspot operators (of unknown abilities, who have never seen the show before and, in worst case theory, may not be very cooperative).

Lighting Technician

The lighting technician is an electrician who does the physical work of assembling the lighting rig, hooking up to house power, focusing the instruments, and maintaining the dimmers. If the local stagehands are unionized but the road technician is not, he will do a lot of pointing, so he must be diplomatic and clear in his instructions.

Sound Mixer

The sound mixer is often quite like the lighting designer. He or she may or may not be involved with assembling the sound system during the setup, but does run the control console. He or she may be an actual studio-trained mixer with many recording credits who has been pirated away by the artist, or may be just a road sound technician with a good ear and a deft touch at the controls who has moved up the ladder.

Sound Technician

Like the lighting technicians, sound technicians do the physical work of assembling and maintaining the sound system.

Equipment Manager(s)

The equipment managers handle the basic band gear, set up the amps, drums, keyboards, et cetera and remain ready during performances to deal with a broken drum head or guitar string and generally aid the band. While lighting and sound personnel may work for companies contracted to the artist, equipment managers are individuals who generally work directly for the band and are not under outside contract to any company.

Stage Manager

Concert tours do not generally have a stage manager as we think of one in theatre. The promoter usually provides a person to act as liaison during the load-in and setup. The head equipment manager, or perhaps the road manager, often does the same for the tour. Only on large or complex shows will there be a need for a tour stage manager who acts as setup supervisor and all-around troubleshooter.

Pyrotechnician

The pyrotechnician is an experienced technician, usually required to be licensed by the state in which the show is playing, who handles flash pots, smoke pots, explosives, and similar effects. Local ordinances are usually very strict about the use of such effects, and although a few acts with especially heavy effects carry their own pyrotechnician, most acts have the promoter hire someone who is locally licensed.

Rigger

The human fly called a *rigger* ascends the heights inside fieldhouses, ice rinks, and anywhere else necessary to secure hanging points for flown lighting and sound systems. The best riggers come from ice shows and circuses, and find that the fast pace and reduced setup time of concert tours offer a unique challenge to their daring abilities.

Audiovisual Specialist

The use of multiscreen slide, film, and video projections is on the increase in rock and roll touring, and this has created a need for a separate specialist who sets up, maintains, and operates the audiovisual equipment on the tour. Almost all outdoor concerts now travel with a video crew to provide large screen video magnification of the performance. Some indoor arena shows have added this element too.

Scenic and Costume Designers

As more rock acts recognize the value of an interesting set, scenic designers are finding more work in rock and roll touring. Durability and ease of assembly are prime considerations for touring sets, and although most sets are fairly simple, the occasional super-act will stage an elaborate production that requires a lot of scenery. Costumers are not currently needed for many artists. One very good area for costume work is with the Las Vegas style performer. Disco, funk, and heavy metal acts also go in for exotic costumes but the general apparel worn on stage by many rock performers is less than a coordinated ensemble.

Laserist

The laserist operates and maintains laser equipment, which is used frequently by tours for special effects. The facility will need to provide some special requirements such as the constant water supply required for high-power lasers. These requirements must be clearly spelled out in a contract rider. Also, local laws may require inspection prior to the performance by a health and safety officer; arrangements for the inspection will take time to set up.

As We Grow: The Question of Unions

The average age of the road technician ranges from 21 to 30, but we are seeing a growing number of old-line legitimate theatre technicians being drawn into touring by the high wages. These workers are often age forty or older; riggers in particular seem to be older men as a result of the training and years of experience required for competency in their field. It is also true that as many superstars are turning forty, so are the long-time touring people who started in the early days and have stayed with it.

Not many rock acts are staffed by union stagehands. Sound and lighting companies that provide road technicians are divided over this issue, and all the time-honored union/anti-union arguments are being bandied about as would be expected. Wages, as always, are at the core of the controversy. The anti-union side feels that they could not afford to pay the union scale, and that the union, once installed in their ranks, could control future wage increases.

As could be expected, promoters, producers, and managers are not very interested in paying additional wages. There are, of course, objections other than wage scales and no one can be sure just how the unionization issue will be ultimately resolved. However, those companies that have gone along with unionization have had ACT (Associated Crafts and Technicians) cards issued to their employees by the IA's International office in New York. Pensions and other privileges are extended to ACT card bearers but not specific membership in a local. Basically, the card allows concert technicians to work in halls with union jurisdiction. Some locals have not taken kindly to the ACT plan, feeling that their members should be getting jobs before persons who do not belong to any local. This attitude varies around the country, of course, from local to local and from East to West. I have had crews work this way and they have generally been treated very well by the local crews.

The United Scenic Artists (USA), the union that generally represents the interests of theatrical designers in the United States, has not made any real effort to get involved specifically with concert touring. The average rock and roll lighting design fee is far above the USA minimum scale. Because many designers also go on the tour and work the actual show, wage arrangements would have to be worked out on an individual basis, as there is no current fixed fee schedule to cover this type of work.

3

Business

It is relatively easy to get a group to say, "Okay, do our lights!" It is not so easy to keep from getting ripped off. One problem is that most designers go into a meeting eager to show the group how much they know, so they spill their creative guts. Do not be so naive as to think the manager is not mentally taking down every concept you throw out, even if verbally he reacts differently. All too often your ideas show up on stage, but you do not!

This is an old story, and because it really does happen, the United Scenic Artists (USA) and similarly the International Alliance of Theatrical Stage Employees (IA) unions have specific rules that no member puts pen to paper or presents an idea to a prospective client until a contract is signed. Excellent rule, but you need two sides to play the game. Rock and roll has only one side, the manager or producer of the artist. The other side does not exist. Sure, the USA and IA would love to have designers work under their banners, but frankly, they did not realize the economic potential to their members. They were not alone; most of the adult world felt that rock and roll was just a fad. A plan geared to the rock and roll designers' needs and the setting of the industry standards is a long way from being a reality.

What are the economics of rock and roll touring? Who makes the money and how much? Should you work for a company, an artist directly, or be independent (free-lance)—which is best? These are questions that you should consider before you walk into that first meeting.

Since there is no governing body setting fee standards in rock and roll, there is a range of fees, which have been static throughout the 1980s because of increased competition. The 1990s do not hold much promise for dramatic changes. The economics are such that it is all relative to how much you are in demand as a designer. The idea is to create the demand so you can get paid whatever you want.

Free-lance

There are three roads to follow to lead you to work. The first, independent or free-lance way, is simple. All you have to do is find a client and convince the artist, road manager, girlfriend, manager, accountant, business affairs manager, and several close friends of the artist that you can do a great job. Realistically, there is a step before this—how do you get past the secretary in the first place? A good test when I was starting out was to try and get one of the top managers on the telephone. If you accomplish this, maybe you should go directly to the presidency of a record company; why stop at being a mere lighting designer!

But when you get a meeting set, the question is: How much do you charge and where are you getting the equipment? It is rare that a group will hire you to design a show without you also estimating a budget for the equipment it will take to do your design. Therefore, you must convince an equipment rental company that you have a client on the line and you need their best price. Cross your fingers the company does not go directly to the artist's manager and cut you out. I am not implying that this is a regular occurrence, but it has happened. Some of the things the rental company will need to know beyond how many dimmers, lamps, and trusses you will want are: length of tour; personnel requirements; equipment list; who covers hotel, trucking, and travel expenses; and deposit and payment method.

The Company Way

Increasingly, managers do not want to be bothered with payments to multiple companies or to a lot of individuals. The *package* approach is very much alive. The company that can supply sound, lights, trucks, and travel arrangements, et cetera is in a very strong position. Often they sell their services like packaged stereos: good speakers, bad amp, okay turntable.

Herein lies one of the reasons for a designer to hook up with a company. Actually, there are several excellent reasons:

1. Accessibility: The company has been at this game awhile and already has clients. They can get you past the first steps outlined for the independent; then you can get down to work.
2. Learning rock and roll: Designing for rock and roll is a new ball game to most college-trained technicians. Although your education is a good base, you still are not ready to design Madonna or Prince. The best way to learn the ropes is to work for a company already in the business. Working for someone else might not be your goal, but it can help you get your feet wet while you get a paid introduction to the field.

You will lose a little freedom when you join a company because there will be rules and procedures, and a boss who may appear to interfere with your creativity, or put limits on the equipment you can have from the company. But most often the boss is just trying to keep costs within the bid given to the manager. However, your management and sales responsibilities are eliminated and this is a big load off a new designer.

Direct Contact

The third method of employment is to work directly for the group or artist. A growing number of artists own their equipment and therefore need production staffs. Working directly for the artist does have advantages, but I feel that in the long run, you can go stale creatively. You could also become involved in the infighting that plagues so many artists' personal organizations. Still, this way you do get a weekly check and if you are into "hanging out" with a star—we call them paid friends . . . it is still my least favorite option for starting in the field.

Pay and Per Diem

Since there are no union guidelines, your income range is really wide. From a survey I sent to fifty concert touring rental companies and free-lance designers in June of 1987, I was able to extrapolate the following figures. If you work for one of the major concert lighting companies, you can expect a road salary of about $350 to $500 per week while on the road, depending on your position on the crew. Back in the shop between tours the pay is usually less. Most people are amazed to learn that the IA road contract only asks for an amount that falls right in line with what the nonunion people are getting, and in many cases even less. Occasionally, a good head electrician, whether union or nonunion, will get in the range of $1,000 a week, but this takes some climbing up the ladder. Concert lighting directors can and do make more, $1,000 to $2,500 a week, but, as I have said, what people say they get and what is the truth is hard to know. In addition to the road pay, a designer will get a flat fee of $5,000 to $10,000 up front for the pre-production meetings and design time he invests in a tour.

You can also expect a minimum of $25 per day as per diem but the average is $35 per day. This is nontaxable income that the government allows for food and expenses, et cetera for each day you spend away from home on business. The current (1988) legal deductible limit is $44 per day plus hotel and travel. Most accountants will advise that if you get $25 or more per day, receipts should be kept. Consult a tax expert as to exactly what you will need to maintain the tax-free status of this money. Yearly tax changes can eliminate many deductions, so keep informed each year in order to take advantage of whatever tax breaks there are.

Besides these monies, your hotel and travel should be covered by your employer or the artist. Some companies and artists try to work deals whereby you get a flat amount per day but you must cover your own hotel. This is a common practice on theatrical bus and truck road shows. I am against this practice for one-night tours. Your day is already too busy to try to make such arrangements, and you are forced to stay at cheap hotels while the band is probably staying at an AAA hotel. I do not believe that the technical crew should be treated like second-class citizens and neither do most artists. It is only accountants and business managers who don't understand the rigors of touring; they are just looking at numbers.

Equipment Costs

I sometimes believe the saying, "There is always someone else who can do it a nickel cheaper!" was originated in rock and roll. When I speak to a prospective client, or even some old clients, this occasionally comes up. There is no question that there is always someone out there who is hungrier. As a designer, you must also be a good salesperson. Your business is your design ability and possibly your crew, and very likely, they can make the difference. You should pay your people as well as you can, so this increases the price you must get for a project. I do not know how to combat this, or frankly whether one should. The kinds of clients who always look for ways to save a nickel here or there are just playing us off against each other. They have no loyalty to anyone and everyone eventually loses.

Like salaries and fees, there are no official guidelines to equipment costs. A company can get $20,000 a night for a tour (the mega-tours do not follow any formula) while another will only get $10,000 for a similar sized system and crew. Although I cannot be sure that what the companies tell is always the truth, and trying to account for the bragging rights factor, the figures given below represent norms set this past decade and are not likely to change well into the 1990s:

4 Genie Tower Systems	$450 to $550 per show
1 Truss, 2 Genie Towers	$650 to $750 per show
2 Trusses (90 lamps)	$850 to $950 per show
2 Trusses (120 lamps)	$1000 to $1300 per show
Truss Grid (150–250 lamps)	$1500 to $2500 per show

Note: All figures are based on a guaranteed five shows per week.

Contracts

Since the USA and IA standard contract forms do not normally apply, you are left to your own devices. I had a $20,000 lesson in contract writing in my early days. I got a group to use my services, so I wrote up what I thought covered all the points of our agreement and a representative of the group signed it. Troubles developed within the group and they broke up. It took over three years to collect my money and lawyers' fees (which had come out of my pocket).

The answer is not necessarily a long legal contract. Part of the problem comes from the extremely fast pace at which this work is done. On an average, a group finally gives you the go-ahead three or four weeks prior to the start of the tour. If you use a lawyer to draw up a contract, you can wait at least a week or two for the finished document to be delivered to you. After the contract is presented to the group, it will go to their lawyers and the process starts over again, which could mean the tour will be over before the contract is signed!

I have found a happy medium with *letters of agreement*. These look rather informal, no "whereas, etc., etc."—just plain language that tells all the details and duties of each of the parties. But even this informal paper must contain the basic components of a legal contract to be valid. The best thing to do is to confer with an attorney who will give you a list of things that must be covered to protect you properly. I have devised a checklist of points that must be covered in all my letters of agreement. Do not use this as a legal reference, because each state has conditions that should be verified by an attorney. But the following will help you understand the basics.

Who Are the Parties Involved?

It is not as easy as it might appear to get into writing the responsible parties. Often the person you have negotiated the agreement with has no legal standing or power to execute a contract on behalf of the party you wish to be responsible for your payment.

What Are You Going to Do for Them?

You must write a job description. Be specific and include even things you assume they should know you would do as a designer, such as provide plans and color charts, call the cues, travel with the show, et cetera.

What Are They Going to Do for You?

You want the client to pay you, of course. But how much, how often, and in cash or check? Are they paying travel expenses? All of these things must be spelled out clearly if you do not want to have it come up later *in court*. Payment schedules that are simple and straightforward are best. Accountants and tour managers like amounts that get paid regularly rather than many add-on charges that cannot be determined before the tour starts. Remember to ask for program or air credit for yourself if shots are used of your show design for a television special or music video. If you also ask to be paid additional fees for such use of your tour design, it will not go over well and could be a deal breaking point. Most managers will not sign such a clause; they feel it puts them in a bad negotiating position. You will most likely get credit, but they will not want it to be locked into a written contract.

What Are You Providing?

You provide yourself, a staff, and equipment. Explain exactly what, in detail, you will give them for their money.

When Will the Tour Begin and End?

You need in writing the starting date of rehearsal, and dates of first and last shows, so that if the tour is cancelled, you have justification for a claim for loss of income.

An Optional Paragraph

Your final paragraph can include the standard legal line about suing the other party if they do not live up to the agreement, and that it is at their sole cost and expense for any legal fees. Actually the expense is set by the Court and does not come close to today's high legal fees. It does show, however, that you do know something about the law and it should keep them on their toes.

Authorized Representative

Finally, have the *authorized representative* sign a copy, date it and make sure the person signs "On behalf of XYZ Productions" and returns a dated copy to you. Since the Uniform Commercial Code has not been adopted in all states, there are many variations and degrees of force with which the courts will hold a Letter of Agreement as legally binding in this simplest of forms.

Consult an attorney in your own state to set up a personal legal plan for your specific situation and needs. The expense may seem high, but not as high as the expense of a lengthy court battle.

If there is a single area in which most designers fail, it is business. Read some books on contract law, business law, and accounting. It can save you great expense and trouble later on. Do not try to be your own attorney or accountant. These are areas where a little knowledge is very dangerous. There is just too much to deal with and it takes years of training to become an expert. Since most people are eager to work, they tend to jump in with both feet before they take a look under the water. Even if you are going to work directly for the artist or the production company, get an agreement in writing that spells out these simple points. You will not be sorry and the time you spend will be worth it.

The Contract Rider

As theatre, film, or television lighting designers, normally we do not become involved in the stagehands' contractual part of the production. In theatre, a union head electrician would be hired for the show and he would assist the union in arranging with the producer for the crew. Concert tours take a slightly different road; virtually none of the shows have the traditional unionized road crews, so there are no department heads to discuss crew needs in the pre-production meeting. You will find that the artist's manager is looking to you to know how long the setup will take and how many local stagehands it will require. Actually, the crewing, physical stage needs, and other items to be supplied in each city in which the concert will play are usually determined jointly by the lighting director, audio engineer, and road manager.

So how do you get your requirements for stagehands known in the different towns? There are two ways, both of which should be used: the *yellow card* and the *contract rider*.

For a concert appearance, the booking agent sends a general contract, which states the flat performance fee or percentage splits of gate, deposit, billing, and other financial conditions, as well as date and time of show, length of performance, et cetera, to the talent buyer or promoter, who signs and returns it. The rider, which covers the specific performance needs of the artist, is usually put together by the road manager after consulting the lighting and sound people who are doing the show.

The *contract rider* is usually considered a supplementary agreement, having its basis in the original contract and being incorporated into the original contract by reference to it. The problem is that if it is not part of the original agreement it is, in effect, a new contract and therefore there must be a separate and distinct passage of consideration from the offeree. What is it that the artist offers the promoter to accept the rider?

The law considers something called *trade usage* or custom. This means that if you can prove in court that it is a widely used and generally common practice in your business to send riders containing certain information and demands for equipment and services, after the original contract has been signed, a court could accept the rider as a valid part of the agreement.

One final note: If you are in a position to send a rider to promoters, send it registered mail, return receipt, so you know that it got to them. At the very least, send it via an overnight package service. People tend to take this form of delivery as more important than a regular airmail letter and will usually see that it gets to the individual as soon as possible.

Follow-Up

The major problem is not what to put into this rider and in what degree of detail, but will it get to the right people in time? This is why a follow-up or hall advance should be made in conjunction with the rider. After making sure that each promoter gets the newest, most accurate rider, it is very important to follow up to see that the promoter gets it to the person whose job it will be to arrange for those things the rider requires. If your artist is not making his or her first headline appearance, then there is already a rider floating around out there. Since it is usually not attached to the contract, it is probably the old one. Good follow-up can take care of this.

Your problems are not over. Chances are that you will get the rider back with changes, deletions, and notations. At that time, check with the booking agency and the artist's management to make sure they are aware of any such changes or deletions. Often, a booking agent receives a modified rider, signs it and sends it back to the promoter, and never tells the road crew. Producing your best effort if the stage is five feet too narrow for the ground-supported truss or if there is only one followspot instead of the four you had planned in the design takes a lot of physical as well as mental adjustment.

Rider Items

There are three general areas that must be covered in the rider:

1. Artist's requirements
   A. Piano, tuned to A440; specify size of piano.
   B. Piano tuner to tune piano prior to sound check and be available after sound check for touch-up.
   C. Amplifiers, organ B-3 with one or two Leslies, electronic keyboards.
   D. Dressing room needs; how many people in each.
   E. Limousines required; airport and hotel pickups.

2. Food
   A. Beer and soft drinks for crew during setup.
   B. Breakfast (if early load-in), lunch, and dinner for crew.
   C. Food trays in dressing rooms for artist; with cheese tray and fruit, and whatever liquor and beer—many artists specify brand names.

3. Stage requirements
   A. Time of load-in.
   B. Number of stagehands required.
   C. Number and type of followspots, possibly position (front, rear, side).
   D. Power requirements for lighting, sound, and band.
   E. Stage size.
   F. Rigging requirements.

In addition to this basic group of items, the rider must include the specifics that pertain to the actual production. These include any items that could incur cost to the promoter or people or services expected to be provided, such as balloons or a seamstress. Special note should be given to alert the promoter if truss-mounted followspots will be used. Some stagehands will not operate them, and a half hour before show is no time to find out someone will not do the job.

The Promoter's View

Interviews with several promoters revealed their feelings about the contract riders they received. Their views were well summed up by people from Concerts West, one of the largest promoters of concerts across the United States. They have done tours for Bad Company, Wings, Neil Diamond, John Denver, Elvis, The Beach Boys, The Moody Blues, The Eagles, and many others. They expressed the promoters' constant annoyance with out-of-date riders, but conceded that it was possible that the booking agent or a management secretary sent out a rider that had been lying around without first checking to see if it was the latest version.

From the promoter's standpoint, essential information that needs to be covered in the rider is: stage size, power requirements, number of followspots, security needs, and band equipment not being provided by the artist. Asked what is most often left off the rider, they specified keyboards, which the artist expects the promoter to rent for the performance, and the request that a piano tuner be there.

Sixty percent of the riders they see are clear and complete, but the remainder are confusingly written or simply are not complete. Many promoters voice complaints about catering requirements. They feel that artists insist on too much food and beverage. Most artists just do not consider the promoter's cost of meeting the rider.

It was also pointed out that although most riders make clear how many stagehands will be needed, they do not break down the time—four hours to set up, two hours for sound checks, three hours to strike and load out the show, for example. Since some union fees have different rates for these periods, it would help if the artist gave the average time required for each part of the day's activities.

In summary, the promoter wants clarity, reasonable requirements, and up-to-date riders. Too often the rider spends more time on additional frills than on services required to put on a good show. Requests for pool tables backstage and limo pickup for girlfriends, et cetera, can mean the difference between the production having that fourth followspot or not.

Although not actually part of the rider, insurance is a big expense for the promoter and the fees are based on the past history of all the shows the promoter has done. That means that the promoter must split the cost between the shows being promoted that year. Some artists have begun to take out their own insurance because of their past good record, thus obtaining lower rates if they have not had problems with equipment or, more importantly, audiences. This is no small sum. On a per show basis, the fee can run $2,000 to $15,000 for liability insurance on a concert. If the promoter does not have to pay this, he has more money to put into the physical production. Thus, the rider benefits.

Small Production, Low Budget

What about the artist who does not carry lights or sound on the road? This is usually a new rock and roll artist, but also includes many Vegas-type acts as well as many jazz and country performers. But these artists are becoming more and more aware of the need for production values, and they are being brought into the rock and roll mode of production.

The rider must be written so that it can be understood by the electrician or promoter's representative who will be given the paperwork to prepare the show. Do not be fooled; the standard theatrical template may not be as universally understood as you have been led to believe. Somehow you must produce a light plot that it is reasonable to believe the promoter and facility can re-create. Just leave enough leeway so that you can be flexible when it comes to the physical limitations of the facility.

If you do not carry the lighting equipment, it is best to give colors in general terms, light red and moonlite blue, rather than specify a #821 Roscolene, because there are several very good color medias available; sometimes not all brands are available in every town. The same holds true for instrumentation. Will you get all PAR-64s when you needed some ellipsoidal spotlight? Try to make clear the area each lamp will cover and its function, so if a substitution has to be made, the local supplier or electrician can give you something that will come as close to your needs as possible. Make sure you specify a working intercom between followspots, house lights, light board, and yourself. Although good intercoms are the rule in the 1980s, I still specify as follows:

The intercom shall be of the proximity-boom-mike and double-headset type and shall talk back at all positions. No hand microphones or telephone operator sets shall be acceptable.

The actual form of the rider need not be drawn up by a lawyer. But good English and clear presentation of your needs are a must (see Figure 3-1).

After the rider is completed, you are only one-third of the way done. The remainder of the work is follow-up. Of that, one-half is tracking down the person who *should* have the rider for each venue, and the other half is explaining and working out the compromises. You could say that the contract rider is a waste of time, if in fact you need to work out all these compromises. But by covering all the bases, you should improve your chances for success.

Ideally, a single promoter will be doing the whole tour (it is happening with more and more frequency) and then you can get everything set once and not worry—much.

Figure 3-1 Sample page from contract rider

4.4 Color media to be placed in all followspots:

Frame #1 819** Red
Frame #2 817 Amber
Frame #3 857 Blue
Frame #4 838 Magenta
Frame #5 842 Lavender
Frame #6 826 Pink

**Colors can be equivalent in other manufacturer's line. The Roscolene numbers are indicated for reference only.

4.5 An 18″ mirror ball and motor to be hung on stage shall be provided by the promoter/producer at his sole cost and expense.

4.6 If the following lighting equipment is not permanently installed in the facility, the equipment is to be provided by Sundance Lighting or their designated subcontractor where practically and economically feasible.

4.7 All stage lighting shall be controlled via electronic dimmers and controller at the sole cost and expense of the promoter/producer.

4.7A A multiscene control desk having a minimum of two scenes and independent functions with scene master control, slaving, and grand master. Ten-scene pin matrix and chase control are desirable.

4.7B Nondim control of mirror ball and on-stage strobes (two line outlets at drums) shall be required.

4.8 Lighting fixtures and position required are as follows:

3 color per side, side light position circuits (lavender, amber, and magenta)

1 color top light circuit on band (blue)

3 color cyc light circuits (red, amber, and blue—green optional)

6 - 6 × 16 Lekos top light circuit at front of stage (no color)

3 - 6 × 16 Lekos as special pool lights (lemon, lavender, and blue)

4 - 6-lamp PAR-64 spot back light circuits (lemon, blue, red, and white)

2 front light wash circuits (lavender and pink)

(A light plot is available from Sundance Lighting Corp.)

The Yellow Card

A *yellow card* show is one that is negotiated with the International Alliance of Theatrical Stage Employees (IA) local, usually in the town in which the show rehearses or previews. They work out the needs for all the road and local crew in advance. The form that is sent to each IA local happens to be yellow, thus the name commonly used in conversation. It specifies how many persons in each department—props, carpentry, electrics (sound is still considered part of the electrical department in theatre)—are required for load-in, setup, show run, and strike. The IA local's business agent then sets the calls for members based on these requirements. The card also shows how many people in each department are traveling with the show. You can be sure the business agent of the local will check to make sure each road person listed on the yellow card has a valid IA card as well as a pink contract (a road contract issued by the local).

A few rock and roll shows have yellow cards, but it is rare. A yellow card cannot be issued unless *all* road personnel are IA members and hold valid cards and pink contracts. Just because someone holds an IA card does not mean he or she can get a road contract—certain requirements must be met. So make sure, if you are in a position to hire, that you ask the applicant if he or she can get a contract to go on the road, not just "Do you have a union card?"

The chances that the rock and roll sound company on the tour has been unionized are even slimmer. None of the major sound companies are currently unionized. This effectively blocks the issuance of the yellow card, even if the lighting companies' employees and the carpenters and riggers are union members. So why even tell you about the yellow card if it is not possible to obtain one?

There are always exceptions to the rules. Besides, I still harbor the hope that tours will become organized. Without going into a long discussion, just remember what I said at the opening of this chapter—you need two sides to play the game and as presently constituted, we only have management. The designer still has no way to find a benchmark by which to judge his rates. That, of course, is only one part of the problem. The real issue is longevity and what happens to the road technician after many years of service. Only a very few artists offer retirement and health plans to their crews.

The Importance of the Rider

Write everything into the rider and be as clear as you can, because if you believe that everyone knows that you need power on stage to do the show—guess again! People with money promote concerts, not necessarily people who know what it takes to mount a production. It can happen that the promoter is extremely naive, and so taken with "show biz" that he thinks everything is magic, including how a show is set up.

The first reason I place so much importance on the rider is that it shows the promoter that competent production personnel are on the show and gives a sense of security that the coming production is together. Confidence is half the battle.

Second, it helps you get your act together. By taking the time to write a clear, full, and accurate rider, you do your homework, and that helps to anticipate problems before you hit the road. There is no substitute for good planning. If there is ever a time that Murphy's

Law will come into play, it is on a concert tour of one-night stands. You cannot avoid problems completely, but you can be better prepared if you have spent the time in the pre-production stage.

4

Pre-Production

Before you can sit down and start a light plot, even for a play, you need to know the physical criteria of the locations. For the concert lighting designer there are some added twists. It is impractical to look over floor plans for the forty or more facilities in which the artist will perform on a tour. And because facilities will vary in width, height, and power availability, it is difficult to design without this information. Where do you get your facts? Usually it is the road manager or, possibly, on a large tour, the production coordinator.

The chances are slim that a city-by-city, hall-by-hall schedule will be available a month or two before the show hits the road; the schedule usually arrives a week or less before the first show, if then. So we must deal in broader classifications, that is, theatres, arenas, college gyms, outdoor festivals, racetracks, et cetera.

A checklist of the basic information needed includes:

1. Type of halls to be played (theatre, arena, outdoor)
2. Budget (per show or weekly, and what it must cover)
3. Artist's requirements
4. Stage limitations
5. Crewing
6. Opening acts
7. Prep time available (pretour and on the road)
8. Rehearsal time available (before tour with lights)
9. Contract rider, as it existed on the last tour (see Chapter 3)

Type of Halls

If the manager can at least narrow it down to the type of halls to be booked, you have the most important piece of information. However, this will be clear to you only after you have played a variety of buildings. You must use your own judgment in limiting your staging, based on general categories like theatre or arena, knowing full well the variables from structure to structure are great even within these groups. The type of performance spaces artists play range from arenas to clubs to rodeos to colleges, and anything in between.

Some reference books are available (see the list below). *Talent and Booking*, which was published irregularly but is now out of print, listed venues (facilities). If you can find an old copy, it will be very helpful. Its main function was to list recording companies, talent managers, sound and light companies, radio stations, colleges, clubs, and promotional firms. *Performance* is a weekly trade publication. Its main editorial policy is to list tour schedules (see Figure 4-1). Each issue also lists upcoming tours (Figure 4-2). Another feature is the spotlight on a road crew (Figure 2-1), shown in Chapter 2. From time to time, it also spotlights facilities, which makes it a good source for learning about the different venues and what they offer in the way of size, et

Figure 4-1 Itineraries
Listing of concert itineraries from *Performance* magazine. (Reprinted from vol. 17, no. 15, p. 31, by permission of the publisher.)

• Itineraries •

DIRECTORY OF MAJOR AGENCIES

WITH OFFICES IN MORE THAN ONE CITY

AGENCY FOR THE PERFORMING ARTS (APA)
New York (212) 582-1500; Los Angeles (213) 273-0744.

WILLARD ALEXANDER, INC.
New York (212) 751-7070; Chicago (312) 236-2460; West Hollywood (213) 278-8220.

EDMONDS TALBERT TALENT CONSULTANTS
Midwest (312) 871-3334; East Coast (201) 273-2090

JIM HALSEY COMPANY
Tulsa (918) 663-3883; Beverly Hills (213) 273-2473; Nashville (615) 329-1700.

IN CONCERT INTERNATIONAL
Nashville (615) 244-9550; Beverly Hills (213) 273-4726; Los Angeles (213) 275-3900

IN TUNE TALENT
Los Angeles (213) 465-9135; Levittown (215) 946-0548

INTERNATIONAL CREATIVE MANAGEMENT (ICM)
New York (212) 556-5600; Los Angeles (213) 550-4000.

BUDDY LEE ATTRACTIONS
Nashville (615) 244-4336; New York (212) 247-5126; Kansas City (816) 454-0839.

MAINSTAGE MANAGEMENT INTERNATIONAL, INC.
Long Beach (213) 433-6771; St. Louis (314) 725-5051; Annapolis (301) 268-5596.

MAJESTIC ARTISTS
East Coast (201) 265-8313; West Coast (213) 460-6393.

WILLIAM MORRIS AGENCY
New York (212) 586-5100; Los Angeles (213) 274-7451; Nashville (615) 385-0310; London 01-734-9361; Rome 868-551-2-3.

PRODUCERS, INC.
Tampa (813) 988-8333; Atlanta (404) 451-5532.

SPOTLITE ENTERPRISES, LTD.
New York (212) 586-6750; Los Angeles (213) 654-5063; Tulsa (918) 582-6750.

STARVISION PRODUCTIONS
Dayton (513) 228-5400; Las Vegas (702) 366-0739

SUTTON ARTISTS CORPORATION
New York (212) 977-4870; Los Angeles (213) 820-8110.

TALENT CONSULTANTS INT'L. (TCI)
New York (212) 582-9661; Los Angeles (213) 460-4209

NORBY WALTERS ASSOCIATES
New York (212) 245-3939; Beverly Hills (213) 275-9449.

PERFORMANCE Itineraries consist of two basic types of listings. Complete Listings include all dates in our data base scheduled for an artist starting with the current issue date. Changes in the itinerary from the last time it was published are printed in bold face type. Any time a current update (even a single date change) is reported in an artist's itinerary, the complete itinerary is published, highlighting the update in bold type. Partial Listings include the name and contact for artists currently on tour as reported to our data base and the issue date of the last time the act's Complete Listing appeared in *PERFORMANCE*.

New Itineraries This Week:

Cabo Frio
Faster Pussycat
Peter Gabriel (Europe)
Hoodoo Gurus
INXS
Manowar/Heathen
Alison Moyet
Righteous Brothers
Scruffy The Cat
Frank Sinatra
Percy Sledge
Steeleye Span
Suicidal Tendencies
T'Pau

BRYAN ADAMS — [A&M]
BA: ICM (212) 556-5600
Opening Act: The Hooters

OK Oklahoma City	**Aug. 22**
KS Wichita, Kansas Coliseum	**Aug. 23**
KS Topeka	**Aug. 24**
TN Memphis, Mid-South Coliseum	**Aug. 27**
AR Little Rock, Barton Coliseum	**Aug. 28**
TX Dallas, Reunion Arena	**Aug. 29**
TX Austin	**Aug. 30**
TN Knoxville	Sept. 1
WV Wheeling	Sept. 2
MA Springfield	Sept. 4
NY Rochester	Sept. 5
PA Allentown	Sept. 6

ALABAMA
BA: Dale Morris & Assoc. (615) 327-3400
Opening Act: Michael Johnson Sept. 18-20, 25; Oct. 9-11, 14, 30-31; Nov. 1, 6-8, 13-15, 20-22, 28-29

IN Indianapolis, Indiana State Fair	Aug. 21
IL Springfield, Illinois State Fair	Aug. 22
IA Des Moines, Iowa State Fair	Aug. 23
MO Sedalia, Missouri State Fair	Aug. 24
ON Toronto, CNE Grandstand	**Aug. 26**

ON Ottawa, Mulson Music CCE Grandstand	Aug. 27
ON Sudbury, Community Arena	**Aug. 28**
MI Charlevoix, Castle Farms	Aug. 29
MI Detroit, Michigan State Fair	Aug. 31
PA Allentown, Fair	Sept. 2
MN St. Paul, Minnesota State Fair	Sept. 4-5
IL DuQuoin, State Fair	Sept. 7
NE Lincoln, Nebraska State Fair	Sept. 10
CA Anaheim	Sept. 12
MI Allegan, County Fair	Sept. 15
PA York, Inter-State Fair	Sept. 17
OH Columbus	Sept. 18
IL Rockford	Sept. 19
IA Cedar Rapids	Sept. 20
AR Fort Smith, Arkansas-Oklahoma District Fair	Sept. 25
KY Lexington	Oct. 9
TN Nashville	Oct. 10
TN Murfreesboro	Oct. 11
ME Portland	Oct. 30
NY Lake Placid	Oct. 31
NY Binghamton, Broome County Arena	Nov. 1
GA Savannah	Nov. 6
NC Charlotte	Nov. 7
NC Chapel Hill	Nov. 8
OH Cincinnati	Nov. 13
WV Charleston	Nov. 14
MD Landover	Nov. 15
TN Memphis	Nov. 20
AR Little Rock	Nov. 21
MS Jackson	Nov. 22
FL Tampa	Nov. 28
FL Hallandale	**Nov. 29**

MOSE ALLISON
BA: The Rosebud Agency (415) 386-3456

WA Lopez, Islander Lopez	Aug. 21-22
WA Seattle, Center Opera House	Aug. 23
MA Cambridge, Nightstage	**Sept. 18**
MO St. Louis, Mississippi Nights	Sept. 23
MO Kansas City, Grand Emporium Saloon	Sept. 24
MO Columbia, Blue Note	Sept. 25
KS Lawrence, The Jazzbus	Sept. 26
OK Tulsa, Sunset Grill	**Sept. 28**
CA Hollywood, Vine Street Bar & Grill	Oct. 22-24
CA Hollywood, Vine Street Bar & Grill	Oct. 28-31
La Jolla	Dec. 2
CA La Jolla	Dec. 3-27

THE GREGG ALLMAN BAND — [Epic]
BA: Variety Artists Int'l. (213) 858-7800
Appearing With: Stevie Ray Vaughan Aug. 12, 15, 20, 23

KY Louisville	Aug. 21
MO St. Louis	Aug. 23
MO Joplin	**Aug. 25**
OK Oklahoma City	Aug. 26
CO Pueblo	Aug. 28

MO Kansas City	**Aug. 29**
IA Des Moines	Aug. 30
CA Lancaster	Sept. 4
WA Seattle	**Sept. 5**
OR Portland	**Sept. 6**
FL Tallahassee	Sept. 26
AR Little Rock	**Oct. 9**

BILL ANDERSON — [Swanee]
BA: World Class Talent (615) 244-1964

TN Nashville	**Aug. 21**
OH Canton	Aug. 22
OH Dover	Aug. 23
TN Hendersonville, Music Village U.S.A.	Sept. 3
VA Woodstock	Sept. 4
WV Wheeling	Sept. 5
WV Huntington	Sept. 6
TN Hendersonville, Music Village U.S.A.	Sept. 10
AL Decatur, Morgan County Fair	Sept. 17
AL Cullman	Sept. 24-25
TN Hendersonville, Music Village U.S.A.	Oct. 1
GA	Oct. 13
TN Hendersonville, Music Village U.S.A.	Oct. 15
MO Branson	**Oct. 16**
MO Branson, Roy Clark Lodge	**Oct. 17**
TN Hendersonville, Music Village U.S.A.	Oct. 22
AL Birmingham, Alabama State Fair Grandstand	**Oct. 24**
TN Hendersonville, Music Village U.S.A.	Oct. 29
OH Mentor	Nov. 1
OH Cleveland, Convention Center Music Hall	**Nov. 21**

JOHN ANDERSON — [Warner Bros.]
BA: Buddy Lee Attractions

TX Fort Worth, Billy Bob's Texas	Aug. 21
MO Branson, Ozark Mountain Theatre	Aug. 22
OK Bixby	Aug. 23
CO Thorton	Aug. 24
CO Denver, Paramount Theatre	**Aug. 25**
NE Lincoln, Pershing Auditorium	Aug. 27
MI Detroit, Michigan State Fair	Aug. 29
OH McConnelsville	Aug. 30
IL DuQuoin, State Fair	**Sept. 4**
TX Terrell, Lee's Silver Fox	**Sept. 11**
MO Eureka, Old Glory Amphitheatre	Sept. 26-27
GA Marietta, Miss Kitty's	Nov. 4
GA Augusta	Nov. 5

LYNN ANDERSON
BA: McFadden & Assoc. (615) 242-1500

GA Loganville	Aug. 22
PA Forksville, Sullivan County Fair	**Sept. 5**
PA Hershey, Hersheypark	Sept. 6
OH New Lyme	**Sept. 7**
NB Fredericton, Coliseum	**Sept. 10**
ME Presque Isle	Sept. 11
ME Bangor, Opera House	Sept. 12

---

cetera. *Performance* is also responsible for the only formal award recognition of tour designers and technicians via their annual Reader's Poll. Billboard Publishing has produced a supplement called *On Tour* that is along the lines of *Talent and Booking* but without artists and managers listed. It is not published on a regular basis. Another publication called *AudArena Stadium Guide* is published annually by Bill-

Figure 4-2 Upcoming Tours
Example of "Upcoming Tours" in *Performance* magazine for the week of August 21, 1987. (Reprinted from vol. 17, no. 15, p. 10, by permission of the publisher.)

· *Upcoming Tours* ·

Fleetwood Mac
BA: CAA
Phone #: 213-277-4545
RC: Warner Bros.
Anticipated dates: Fall

Loverboy
BA: CAA
Phone #: 213-277-4545
PM: Bruce Allen Talent
Phone #: 604-688-7274
RC: Columbia
Anticipated dates: Fall

Love and Rockets
BA: Triad Artists
PM: Raymond Coffer (UK)
Phone #: 441-950-5489
RC: Beggars Banquet
Anticipated dates: Nov.-Dec.

Danny Wilson
BA: CAA
Phone #: 213-277-4545
PM: Gerald Grimes
Phone #: 443-824-55721 (Australia)
RC: Virgin
Anticipated dates: Fall

Midnight Oil
BA: CAA
Phone #: 213-277-4545
PM: Gary Morris
Phone #: 2-660-5000 (Australia)
RC: Columbia
Anticipated dates: Fall

The Legendary British Blues Tour
featuring Long John Baldry, Spencer Davis, Savoy Brown
BA: In Tune Talent
Phone #: 213-465-9135
Anticipated dates: Sept. 13-late Oct.

Sting
BA: FBI
Phone #: 212-246-1505
PM: Miles Copeland
Phone #: 213-938-5186
Anticipated dates: World tour begins Nov. (T)

Steeleye Span
BA: Producers, Inc.
Phone #: 813-988-8333
Anticipated dates: May (3 wks)

Pentangle
BA: Producers, Inc.
Phone #: 813-988-8333
Anticipated dates: late Oct.-Mid-Nov.

Tears For Fears
BA: ICM
Phone #: 212-556-5600
PM: Paul King
Phone #: 011-441-437-2777
RC: Mercury
Anticipated dates: Jan./Feb. (U.S.)

Tirez Tirez
BA: FBI
Phone #: 212-246-1505
PM: Club Soda Music
Phone #: 212-757-9462
RC: PMRC
Anticipated dates: Sept.-Nov.

Holly Near
BA: Tour Consultants
Phone #: 201-783-0778
RC: Redwood
Anticipated dates: Fall

a-ha
BA: William Morris Agency
Phone #: 212-586-5100
PM: Mel Bush
Phone #: 01-225-1191
RC: Warner Bros.
Anticipated dates: Europe-Dec.; U.S.-Jan.

Jane's Addiction
BA: Triad Artists
Phone #: 213-556-2727
PM: Triple X Management
Phone #: 213-871-2395
RC: Warner Bros.
Anticipated dates: Sept.-Oct. (CA, WA, OR, Canada, NM, AZ, NV, TX, ID)

Cissie Lynn
BA: Music Unlimited (Canada only)
Phone #: 604-658-1313
PM: Doc Holliday
Phone #: 804-838-9552
Anticipated dates: Jan.-Mar.

Tom Verlaine
BA: FBI
Phone #: 212-246-1505
PM: John Reid
Phone #: 011-441-437-2777
RC: I.R.S.
Anticipated dates: Fall

Mr. Mister
BA: Triad Artists
Phone #: 213-556-2727
PM: George Ghiz
Phone #: 213-278-8877
Anticipated dates: Europe/Pacific - Nov.; U.S. - January

Vassar Clements
BA: Producers, Inc.
Phone #: 813-988-8333
Anticipated dates: Nov.

board. It does not give as much technical detail as it does staff, costs of rentals, and services available. It is only one of several reference guides that Billboard publishes throughout the year.

One helpful organization, the International Association of Auditorium Managers, provides a guide to venues—the *IAAM Journal*. It is probably the most complete listing of stadiums, auditoriums, arenas, and theatres, but is designed primarily to provide a listing of hall and facility contacts for ticket sales, catering, and building services. While lacking detailed technical data, it does give hall type, floor or stage size, seating capacity, and other general details, as well as names and telephone contacts at the venues.

The legitimate theatres have a degree of uniformity, because road companies of Broadway plays and musicals are most often booked, but this uniformity is limited to some thirty or forty houses across the country. A concert tour plays everything from an old movie house to a symphony hall and it is still called a theatre tour. A theatre tour for a rock band could go from the Capital Theatre (an old movie house) in Passaic, New Jersey, to the Minneapolis Symphony Hall (no overhead pipes), to the Arie Crown Theatre, Chicago (very deep stage with excellent grid system), to the Constitution Hall in Washington, D.C. (a domed stage, no grid), to Pine Knob Music Center outside

Detroit (another covered stage with open air lawn seating). This is why trusses and other structures are trouped by concert artists even when doing a theatre tour.

When we talk about arena tours, we can expect even less in the way of theatrical facilities, that is, pipes and stage. Arenas are for the most part simply large airplane hangars with seating.

At a few exceptionally well-equipped facilities around the country, the house staff cannot understand why we do not use their fixtures and counterweight systems. What they fail to understand is that the elements of timing, consistency, and repeatability are the keys to a good concert. A theatre production usually rehearses in each new town, so there is time to make adjustments in hanging positions, focus, dimmer levels, et cetera. A touring concert has less than twelve hours from load-in to curtain. The concert designer, as well as the artist, must be assured of what will be seen during the performance.

Publications that List Venue Facts

AudArena Stadium Guide	Billboard Publishing
	Box 24970
	Nashville, TN 37202
Official Talent & Booking Directory (Out of publication)	MPA International San Francisco, CA
On Tour	Billboard Publishing
	9000 Sunset Blvd.
	Los Angeles, CA 90069
Performance	Performance Inc.
	2929 Cullen Street
	Ft. Worth, TX 76107
IAAM Journal	International Association of Auditorium Managers
	500 No. Michigan Ave.
	Chicago, IL 60611

Budget

Some designers refuse to let budget restrictions encumber their creativity. They prefer to get as much information as is available and do their design as they feel it should look. They then present it to the artist and try to convince management that the idea is worth whatever the expense.

Maybe I am too much of a realist, but I like knowing at least the approximate budget. The designers who work out the budget after the design is submitted are, on the whole, the name designers. They have sold themselves with the understanding that they will have a "blank check." So let us consider the other 99.5% of us who must justify our cost to management.

It is wise to ask for a budget when you first talk about the project. Likely as not, you will get blank looks because it has not even been considered, or they will throw it back to you in an attempt to see if you will give them figures lower than they paid last time. One way to get a rough figure is to give a range per show, and see what the manager's reaction is. This is not to say that if you go back with a terrific idea that will cost more than the budget, you will not be able to talk them into doing it. The budget should just be a median point to be used as a guide; you can come in under budget or even over budget if you feel it is worth the effort.

Artist's Requirements

The initial meeting should be with the artist. The only way to get a feeling for how the artists want to look is through meeting them, not the management. You will be creating something that will affect their performance significantly, and you must be in tune with them. There are also concerns, like stage movement and placement of band equipment, to be discussed.

One of my first clients was Billy Preston. I was called in at the last minute to do his tour. At the time I only met his road manager, not Billy. The road manager assured me he had been with the artist so long that he knew his every move and requirement. Naturally, the first show looked like I had designed the lighting for Joan Baez and the design was for Kiss!

A rapport with the artist must be established. Confidence in you as another artist who wants to make the singer appear in the best light (no pun) is essential. It is the theatrical equivalent to the theatre designer's relationship with the director.

Stage Limitations

Many physical limitations must be considered in the mounting of the show. Ask yourself questions such as, is there a backdrop being used that must be lit? In arenas, will there be seating in the rear? In outdoor shows, will the wind factor preclude backdrops and some scenic devices? And what about the sightlines and width of the stage? Because you cannot control the performance spaces in so many locations, the design must anticipate physical staging problems before they develop. Once on the road, it is very hard to keep changing things to meet the day-to-day problems. Time is a constant devil, so preplanning and forethought as to possible solutions to every possible staging question must be considered.

Crewing

Will you hire the crew personally or be contracting a lighting company for personnel and equipment? How many technicians will be on the lighting crew? How much time do they have to set up? How many local stagehands will there be to help out? If required, are riggers to be provided by the promoter or are you taking a head rigger or a full rigging crew with you on tour?

Crewing is vital. You can design the most spectacular show in the world and it will be wasted if it is not ready by 8:00 PM. Remember that time is the unrelenting demon of touring. A touring lighting crew of two technicians is normal; three or four can be found on large shows. They usually do the lighting with the assistance of four to six local stagehands. However, a large number of special effects will take time to set up and can slow things down, as will heavy scenic elements. Who handles the road supervision and repair of the set, scrims, drapery, and such? A road carpenter is still rare on a rock and roll tour.

As the designer you will most likely be responsible for all of these elements. That is not to say that you personally will have to take hammer and nail to a broken flat, but you will probably need to make sure everything is kept in working order.

Opening Acts

The opening act is slighted most of the time. Little consideration is made as to stage space or lighting in the majority of cases. The headline band's equipment people usually refuse to move one piece of their stage gear. This is often out of spitefulness, because it was done to them before they became headliners. But it is a fact of the business that you must deal with.

If the opening act is with the same management company, it will probably get more consideration, but not always. Stars' egos can be enormous and they quickly forget how it was when they were just beginning. Do not take for granted that the opening act will be given carte blanche to add special lighting even if they pay for it. In fact, do not agree to allow any equipment to be added to your rig until you have had it cleared with management.

If the opening act is permitted to do anything, you may still have restrictions placed on the extent to which you can utilize your design.

There is a growing tendency to package a single opening act to play the whole tour. In theatre we look at the show and how its parts are integrated into the whole. In concert touring, 99 percent of the time we design for *one* element of the show only. Keep the perspective of whom you work for and do not try to be a good guy to the other artists, as you could lose your job for that effort.

Prep Time

The first consideration in prep time is the time prior to the tour that is allotted to design and physically put the equipment together, check it out, and be ready to rehearse and hit the road. If you are going to use one of the companies in the business of supplying tour lighting, the time required can be cut to a couple of weeks, or in a pinch much less, because they have much of the equipment already packaged and are geared to prep it quickly. A major delaying factor comes when the trussing must be fabricated. If time is tight, you will be doing yourself a favor by checking out what truss plans can be done with in-stock pieces. Also, remember that how busy the shop is will be a determining factor in prep time and very possibly in cost.

The second area to consider for prep time is the time that will be allocated each day to load in and prepare for the sound check. A show that takes a lot of time to load in, hang, and focus must be planned well in advance. This is so the promoter can provide adequate assistance. Local stagehands and access to the hall for as many hours as it will take to accomplish all these things must be arranged for. Pre-planning has taken on more significance with the new mega-tours, like the Jacksons, Madonna, and David Bowie, which take two to three days to load in.

Ability to judge the setup time is an acquired skill. You can consult with the lighting company in the early stages to talk over your concepts and let them give you their ideas as to crew and setup requirements. Also, remember that it takes time to get equipment from one city to the next. A show that is booked with 600-mile jumps in between will never make a 9:00 A.M. stage call. Figure on one hour driving time per 50 miles of travel, but areas of the country where superhighways are not the rule will lower the speed, as will travel through large cities to get to the venue (such as Madison Square Garden in New York City). The time of year, for instance winter in the Midwest, will be a big factor, too. Unfortunately, booking agents are notorious for their complete lack of consideration on this point.

Rehearsal Time

A few top acts stage a full production (Prince and David Bowie, for example) and will approach the tour as a "real" show. The more rehearsal time available, the tighter the first show will be. However, a long rehearsal with lights is rare. The reality is that the band will rehearse in a small room for a week or two, at which time you should be finding places in the music (solos, accents, stage movement, and so on) for lighting changes. But the availability of large stages and the very high cost to the artist of a lengthy full rehearsal places limitations on this phase of the preparation. Here is where your preplanning and ability to make quick decisions are put to the test. Rock bands are not actors who know how to "freeze" while a mood is set and then pick up the action again. They will play through the song, possibly repeat it if they wish, but I cannot stress enough that a true rehearsal such as theatre companies enjoy is nonexistent in a touring rock show.

Variety of Venues and Artistic Styles

A career in concert lighting does not limit you to one type of music. A great variety of artistic styles are performing in concert venues throughout the country each day. There were over 600 musical artists and bands on the road the first week of September 1988, according to *Performance* magazine's listings. This is a relatively inactive time of the year. The peak seasons are April–May, July–August, and October–November. If you break down these 600-plus tours, only four to six are superstars. About forty of these acts comprise the extra large tours and about 50 to 100 use touring lights and sound. Another 150 to 200 are just getting a foothold and may take out a designer and rent equipment locally or regionally. Another thirty or forty are casino circuit acts playing Las Vegas, Reno, Lake Tahoe, Atlantic City, and some large clubs. This circuit can provide for some very creative designing, so do not brush off Las Vegas or Atlantic City. The facilities are extremely well equipped and they are always looking for something new. The rest of the touring acts are playing as opening acts or simply cannot afford or do not care about lights and sound.

The range of places that are used as performance areas is vast and creates wonderful and yet frightening challenges for the designer. Once the foregoing areas are covered—type of hall, budget, artist's needs, staging, crew, opening act, prep time, and rehearsal time—you can sit down, listen to the music, and start to rough out a light plot. But the frustration created by designing before you know the majority of this information is utterly debilitating. Learn as many facts as possible and be positive in your concepts but be prepared for changes. That is a known—there *will* be changes!

5

The Design Stage

Your first concert design will probably be for a local promoter, a college, or a local band. This stage of building a reputation, and more importantly, building confidence in yourself, usually puts you into the type of show where the artist has no touring equipment and probably has a very sketchy rider that gives a stage plan (probably nothing like what he is currently touring), and maybe a basic color chart. You may not even get to "call" the headliner, only the opening act.

So what are you to do? First, go to the promoter and find out how much can be spent on equipment rental. Second, consider the facility in which you will do the show. Are there the required (or hoped for) number of followspots? If not, it will be a big chunk out of your budget to rent them and pay the operators.

The best design for a show like this is a very general one; it may also be the safest, but that is secondary. Three or four colors for each of the back light and side light circuits may seem too easy, but that is what 80 percent of the local concerts get. If you are sure the star will stay at the microphone, add some back light specials to separate and bring him out from the band. (Later I will show a basic plan for most back truss and side Genie Tower–type lighting rigs that would tour with a jazz or country artist, even many rock bands. I have seen many such simple plans on the road that are used very creatively.)

The truth is, if you pick colors correctly, you can mix and get just about any color you desire. The next stage is to add specials as needed, money or dimming permitting. Too often designers get so hung up with BIG that they forget clarity in design. In addition, a wide range of special effects and projections could be added, but these are beyond the type of show we are discussing here.

Fixtures

Your first consideration for light fixtures is general coverage. The PAR-64 (Parabolic Aluminum Reflector) type medium flood lamp is best for general washes. The circuit you wish to have a tighter focus will be the same PAR-64, but with a narrow or very narrow spot lamp. The PAR-64, and the less used but valuable family of PAR fixtures (see Figure 5-1), are the workhorses of concert lighting. Being relatively trouble-free and devoid of parts to jam or break, the units have the most effective, dimmable light source available among the standard theatrical/film fixtures.

The efficiency and type of beam spreads available for the PAR-64 unit should be viewed in a couple of ways. The reason there seems to be a preoccupation with efficiency is that power availability is often a problem. If you are going into an old theatre or college gym, access to auxiliary power can be a big headache. Consequently, you want as much light as possible for as little power draw as possible. The PAR fits this bill best of all. The sealed beam lamps offer a range of beam

Figure 5-1 PAR family of fixtures
The Parabolic Aluminum Reflector family of lamps ranges in size from a PAR-16 to the more familiar PAR-64. (Photo by James Thomas Engineering Ltd.)

spreads that will give coverage from very narrow to wide flood fields. These lamps can give you concentrated beams of light that will project even the most dense color media. For example, we can compare the beam spread at a 30-foot throw between the very narrow lamp—3.7 × 6.4 feet—and the medium flood lamp—6.5 × 15.5—as measured at 50 percent of intensity. While on the subject, those same 1 kW lamps produce 560 fc and 150 fc respectively at that distance, compared to a 1 kW Fresnel, which could only produce 195 fc at spot and 40 fc at flood focus at the same distance.

The real lighting effect for many concertgoers is seeing the colored light beams stabbing toward the stage. Often referred to as *air light*, these beam patterns are created simply as a design element and may not even be focused on the artist. They become a big part of the overall design of many tours. Sometimes these patterns are more interesting than the artist's music.

Other available choices are beam projectors, ellipsoidals (leko), and Fresnels, besides the various types of striplights. Recently the computer-controlled moving lights were introduced and have produced a major impact on concert design (see Chapter 13).

The beam projector, also an open-faced fixture, does just about the same job as the PAR-64, but with a very low efficiency factor. Beam projectors do get used, but have a bad reputation for falling apart on the road (no matter who the manufacturer).

The ellipsoidal has not been given the status it enjoys in the theatre, but more and more it is being employed for concert lighting. Some people consider me this fixture's biggest booster. I have done concerts with well over 180 ellipsoidals and only four PAR-64 fixtures (see Chapter 14). There is no doubt that you will experience higher maintenance factors if you use them, but the tight control and ability to shape the light is a highly desirable quality for more and more designs. However, this is true only when you have a show in which you are confident the talent will be on the rehearsed marks. Artists who cannot hit marks do more than hurt your design; they waste their money. If the effect cannot be seen because of their inability to be consistent in their movements, then the fixtures are not needed and are just so much excess baggage.

The Fresnel sees little favor in the concert market as its weight and limited efficiency do not lend it to this media. Other fixtures such as the film Mini-Brute, an early application of the PAR lamp designed for broad fill lighting for film and location television (see Figure 5-2), and HMI lamps do find special uses from time to time. Generally you will find that the concert manufacturers have produced their own designs for grouping PAR-36 in a single fixture (see Figure 5-3). They

Figure 5-2 Mini-Brute
Either six or nine PAR-36 or PAR-64 lamps can be grouped in a single unit. (Photo by Lee Colortran Inc.)

are usually used as audience lights. Virtually any fixture can be used if the designer has the vision to see what effect it will create and uses it accordingly. But always keep in mind that it must work night after night. Nothing is worse than an effect the artist comes to expect—not working.

Placement of Fixtures

There is a reversal of basic light direction in concert lighting as opposed to theatrical lighting. Whereas theatre puts great importance on front light and least on back light, concerts completely reverse this concept. Two factors are responsible for this. First, there is little or no possibility of consistently getting a balcony or front-of-house position to use for a concert tour in thirty or forty *found spaces* (spaces used but not specifically designed as performance areas). Second, the lighting in concerts is viewed as effect and accent, rather than for the theatrical functions of visibility and mood. Therefore, the idea is to produce all the front light with one or two followspots and concentrate the fixed lighting effects as back light to surround the singer or players with color.

This is known as *accent* lighting to distinguish it from the theatrical style. Light is concentrated at strategic points on the stage to punctuate the music with heavy colors. This method of lighting truly fits the maxim: It's not where you put the light, it's where you don't put the light that counts.

Used as an objective for your designing, this will serve you well to keep you from overdoing the design. Too many shows load in two hundred fixtures when a hundred will do. I had a problem with a group that wanted me to add more lights to the tour because they were cutting back on the set. Their logic was that if they had less scenery, more lights could fill in the void—wrong. You cannot light what is not there, except when using air light! I already had an adequate plot and didn't desire anything more except effects and projec-

Figure 5-3 PAR-36 Spot Bank
This adaptation of the Mini-Brute idea was designed with concert lighting in mind, in groups of four or eight lamps for easy dimming. (Photo by James Thomas Engineering Ltd.)

tions. They had just eliminated the things I needed as surfaces for projection.

The quantity and placement of fixtures must be directly related to the placement of band and vocals. It is unreasonable for a manager to request 80 lamps, or 200 lamps, for his client; my reply is, "Okay, I will use that for the budget figure, but I must know more about the act before I can say if that is a correct number of fixtures for the artist's needs." I see no point in renting 80 (or 200) lamps and then trying to figure out what to do with them, although I admit to being put in this position on several occasions.

One of the most effective lighting designs I have seen was Bruce Springsteen's tour in 1980. His designer at that time, Mark Brickman, had at best 60 lights, but he used them very effectively. I am sure that to the audience, it appeared as a much bigger system. Good design in any field is not based on quantity.

Color

I think in terms of colors, not quantity of fixtures, before I begin to draw a plot. What I am trying to picture in my mind are *looks*, planned patterns of light and color that will be used one or more times during the show. That is a television term and most designers avoid it, but it really is the best way to describe our concerns once the focus of the lighting and solo spots is finished.

Primary and secondary colors are generally best. For a general plot, try for four colors as side light and five colors plus white for back light on the lead artist. Bands usually get four colors for back washes and three colors for side light, plus two back specials—one warm and one cool color, such as Lee filters #105 (orange) and #137 (special lavender) — to accent their solos. I prefer not to use a followspot every time a short guitar break is done. A nice hot back light, while dousing the followspot on the singer, can be very effective. When I am able to get front specials into the design, I add a warm special and a cool special to each solo player, plus at least two general front or top band washes.

Circuiting and Dimming

The fixture complement is tied to the dimming available. It is fashionable not to concern yourself with such trivial matters. After all, you were probably taught to leave that to your master electrician or gaffer. However, reality dictates that availability of equipment most often forces you to pick from a choice of dimmer packages, usually in groups of six channels of a particular wattage: 2.4 kW or 6 kW. Not many 3.6-kW, 8-kW, or 12-kW dimmers are found on the road. The 1-kW dimmers are usually packaged in larger groups, a common quantity being 72 per rack.

Always leave some dimmers with unused capacity so that if and when you have a failure you can gang channels. Better yet, leave two of your largest capacity dimmers as spares in the first place.

Layering

Besides basic position and angle, I look to another idea for most of my visuals: *layering*. This is the process that creates depth and separation by using different shades or saturations of a single color. Except for effects, I rarely bathe the stage in a single color. That gives no visual depth to the stage. By a conscious consideration of color, coupled with intensity, you not only direct the viewers to the part of the stage you wish them to focus on, but you put the total stage into perspective. Too often I see shows that rely totally on the followspot, with its dramatic shaft of bright light, to accomplish this task. That is the easy, boring way out. Not only is the source of light important, but also the hue and tone of the colors used on stage that create a good design.

It seems so simple, yet many people read this and think it means putting a lot of different colors on the stage all at once. You can layer in a single color by hue variation. And do not forget that the absence of light also contributes to this theory.

Where I first realized how important this idea can be was on my early television lighting assignments. In video, the camera cannot show depth; lighting directors must accentuate depth via back light and intensity between the foreground and the background. That same approach can be applied to concerts, where some people are 150 feet away from the stage. In theatre it has application when used to subtly draw attention to a particular part of the stage. There again, not just the equipment you use, but the mix of colors you use can make a better, clearer design.

Layout and Symbols

Lest anyone think that with the drawing of the light plot the design is finished, let me talk about the next step in my approach to design. I have often marveled at beautifully drawn light plots only to be disappointed in the execution of the show. On the other hand, I have often seen light plots scribbled on graph paper or the proverbial envelope that are tastefully executed. However, I will not defend this method even when circumstances have made proper preparation impossible. The effort and wasted time spent in explanation that such a crudely drawn plot necessitates cannot be defended. The designer owes it to the technicians to give clearly understood and accurate instructions and that is best conveyed by the properly drawn light plot.

As you study theatrical lighting, it will become apparent that there are several schools of technique. Some styles are created by teachers at a particular institution of learning, or you will be taught the so-called Broadway method perpetuated by the United Scenic Artists union examination. Generally, the Broadway style prevails, but some West Coast designers have made changes that reflect newer thinking. (Keep in mind that if you do wish to practice your craft on Broadway you will need to learn it as required for the exam.)

Add in the distinctive style of the British school of design that has been injected into the United States and you have an evolving system that no one will agree upon. The United States Institute for Theatre Technology (USITT) took years to agree on standard symbols for lighting instruments; even so, these are still not widely seen on the concert designer's plots.

The concert designers have a style all their own. Partly because a large portion of them are not formally trained and because of a very heavy British influence, they tend to simplify the plot so that anyone can understand the color, circuit, and control channels at a glance.

The major differences between the two styles of drawing (the traditional versus the modern) comprise three areas:

1. Templates: The traditional method uses both top and side silhouettes while the concert designer uses only top silhouettes or even simple circles.
2. Electrical Hookup Chart: Traditionally it is similar to electrical engineering drawings where a line is drawn joining all fixtures of a circuit. Concert designers tend to use a number inside a symbol to represent the patching.
3. Presentation: The East Coast or more traditional method uses separate sheets of paper for many different details needed to complete the design, whereas the concert and Las Vegas approach is to have all the information on one sheet.

The light plot shown in Figure 5-4 has all the information needed to complete the color and hanging of the show. Adapting a tour plot to a Las Vegas venue, such as this one for the Aladdin Hotel, will generally allow for greater individual lamp control and variety of mounting positions than are available on tour. The house fly system allows for the addition of drapery and set pieces not always practical for the road. You can see from the lamp symbols (in the key) that the types of fixtures used are a combination of concert PAR-64s and straight theatre fixtures. I feel this works best in this environment. Because the light plot in Figure 5-4 was originally drawn at a scale of ½ inch to 1 foot there was no room for the circuit chart to be on the same page. The graph sheet (Figure 5-5) makes it easy for the electrician to patch the dimmers and assign the control channels. Whenever possible I have the light plot and circuit chart appear on a single sheet.

Over the years I have evolved a method I believe is even clearer. I have always had trouble printing numbers clearly on drawings, especially when trying to get the color number, circuit number, and control channel number all in the same space. The idea of adding more symbols to the drawing is founded on an idea used by Len Rader, for many years head electrician at the MGM Grand Hotel in Las Vegas. Every light plot that came to him was different, and he said a lot of time was wasted trying to get the crew to understand what the plot was trying to represent. So he started redrawing the plots and keeping them on file so that when the artist returned the load-in was much simpler. On the one sheet he had the circuits, dimmer assignments, and color in a form that everyone on his crew could understand. I adapted his form (see Figure 5-6) and find it easy to understand, having used it all over the world with great success.

Note in Figure 5-6 that the rectangle (symbol for circuit number) behind the fixture symbol is blank. Existing house circuits will be assigned by the staff electrician to best facilitate the hang in the particular house. Note also that fixtures are pointed in the desired direction of focus rather than straight side-by-side as in the more traditional method. The five-foot marks allow for a quick reference to fixture placement along the pipe.

Hang

The final step in your design is not deciding whether to mount the fixtures on a pipe or on a prerigged truss. What looked good on paper may be junk in the air. As the designer you must not only work closely

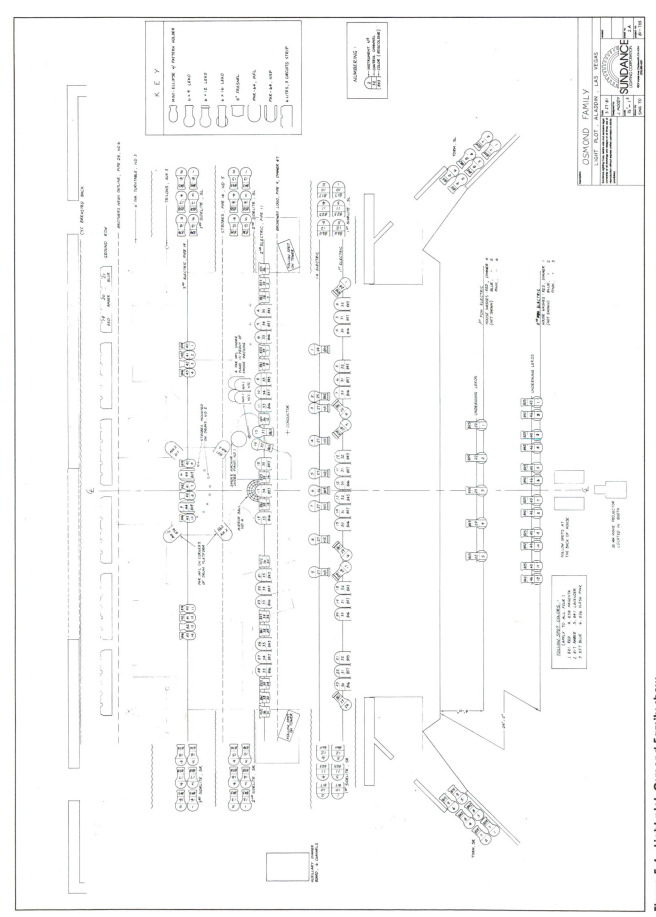

Figure 5-4 Light plot, Osmond Family show
(Designed by James L. Moody.)

Left table (Dimmers 1–35)

Nº	CAP	TYPE	AMT	NUMBERS	LOCATION	DESCRIPTION	FOCUS	COLOR DESCRIPTION	Nº	NOTES
1		HOUSE FIXTURE			2ND FOH ELECTRIC	HOUSE WASH	WASH	RED		
2								BLUE		
3								PINK		
4								RED		
5					1ST FOH ELECTRIC			BLUE		
6								PINK		
7	4K	6x9 LEKO	4	1,2/1,2	TORM SR/SL	SIDELITE		BRIGHT ROSE	829	
8			″	3,4/3,4				SP LAVENDER	842	
9			″	5,6/5,6				GOLDEN AMBER	815	
10		PNK MFL	4	1,2/1,2	1ST SIDELITE SR/SL			DARK AMBER	817	
11			″	3,4/3,4				MED RED	823	
12			″	5,6/5,6				BRIGHT BLUE	840	
13	4K	6x9 LEKO	4	1,2/1,2	2ND SIDELITE SR/SL			MED LAVENDER	843	
14			″	3,4/3,4				LT GREEN-BLUE	856	
15			″	5,6/5,6				MED GREEN	874	
16		6x9 LEKO	4	1,2/1,2	3RD SIDELITE SR/SL			MED LAVENDER	843	
17			″	3,4/3,4				LT GREEN-BLUE	856	
18			″	5,6/5,6				MED GREEN	874	
19	6K	STRIPLITE	6		GROUND ROW	CYC WASH	CYC	RED		
20								AMBER		
21								BLUE		
22	4K	6x16 LEKO	4	1,2,4,5	1ST FOH ELECTRIC	FRONT LITE	BROTHERS	N/C PINK	825	
23	1K	″	1	3		″	DONNY		825	
24	1K	6x12 LEKO	1	2	1A ELECTRIC	BACKLITE	BROADWAY SP	MED BLUE	857	
25			1	6			DONNY, MEMORY	SP LAVENDER	842	
26							MARIE SPECIAL	STRAW	809	
27		6x12 LEKO	6	3,4,5,7,8,9	1ST ELECTRIC		FAMILY	N/C		
28	3K		3	1,9,17		CROSSING WASH	CHIFFON DRAPE	DK MAGENTA	838	
29			″	8,16,24		″		SURPRISE BLUE	861	
30	6K	PNK MFL	6	4,7,12,15,20,23		BACK WASH	DOWNSTAGE	MED LEMON	806	
31			″	3,6,11,14,19,22				MED BLUE	857	
32			″	2,5,10,14,18,21				MED LEMON	843	
33	5K	PNK MFL	5	6,11,17,22,27	2ND ELECTRIC		MID STAGE	MED LEMON	806	
34			″	5,10,17,22,26				MED BLUE	857	
35			″	4,9,15,21,26				MED LAVENDER	843	

Right table (Dimmers 36–47, AUX, ND)

Nº	CAP	TYPE	AMT	NUMBERS	LOCATION	DESCRIPTION	FOCUS	COLOR DESCRIPTION	Nº	NOTES
36	4K	6" FRESNEL	4	1,12,20,31	2ND ELECTRIC	FRONT LITE	COLUMN	N/C	-	
37	2K	6x9 LEKO	2	13,14		BACK LITE	CONDUCTOR	MED BLUE	863	
38	5K	8" FRESNEL	5	3,8,18,25,30		FRONT WASH	ORCHESTRA	SURPRISE BLUE	861	
39	″	″	″	2,7,16,24,29		″	″	MED RED	823	
40	15K	MINI ELLIPSE	3	1,5,11	3RD ELECTRIC	PATTERNS	CYC	MED GREEN	874	#235 - RADIAL LINES
41		″	3	2,7,12				N/C	-	#232 - STYLISED STARS
42				3,8,13				ASSORTED RED GREEN AMBER		#270 - FIREWORKS
43				4,10,14				MED LEMON	806	#302 - MUSICAL NOTES
44	2K	PNK NSP	2	6,9		DOWN LITE	DRUMS	STRAW	809	
45	6K	6x16 LEKO	6	1,3,5,7,9,11	2ND FOH ELECTRIC	FRONT LITE	FAMILY	N/C PINK	825	
46		BROADWAY LOKO	6	2,4,6,8,10,12	PIPE 9	SPECIAL		SP LAVENDER	842	
47										
AUX 1	4K	PNK MFL	4		UNDER SL PIANO	SPECIAL	DRUMMER	N/C	-	
AUX 2	4K	PNK MFL	4		ON DRUM PLATFORM	″		SEE NOTES		863, 821, 817, 874
AUX 3		TRILON LITES			TURNTABLES (4)	″				
AUX 4		TRILON LITES			″	″				
AUX 5		TRILON LITES			″	″				
ND 1		SMOKE MACHINE			UNDER SL PIANO					
ND 2		DRUM STROBES			DRUMS					
ND 3		TRILON MOTORS			UPSTAGE					
ND 4		MIRROR BALL			CENTER MIDSTAGE					
ND 5		STROBE CURTAIN			PIPE 16					
ND 6		BROTHERS NEON			PIPE 24					

Figure 5-5 Circuit chart, Osmond Family show
(Designed by James L. Moody.)

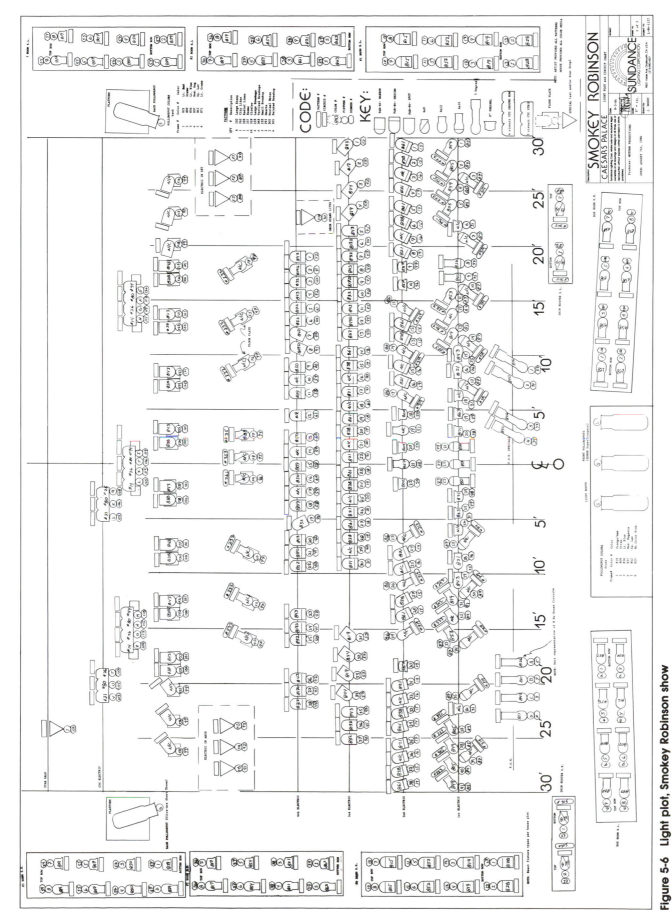

Figure 5-6 Light plot, Smokey Robinson show
(Designed by James L. Moody.)

with the supplier to ensure accurate reproduction of the design, you must also be open to their suggestions concerning changes that will help the repeatability, packaging, and focus of the touring system. If it is a onetime concert, the access to fixtures can be a problem due to scenery or band placement. It might be wise to use a truss so that fixtures can be focused by someone walking or crawling on them. Restrictions as to circuit availability, drapery obstruction, and trim height must all be considered. I never leave the theatre during load-in on a one-nighter. I feel the designer does a great disservice to the crew and shows a great lack of respect for their efforts on behalf of the production when he or she is not attentive during the rigging process.

Often the house stagehands will have suggestions that will help simplify the rigging and focus. It is like hunting or fishing: the locals know the woods and best fishing holes. The designer must keep an open mind to suggestions and not become defensive. Not only does it help to have the crew feel part of the show by your listening to their ideas, it shows your good sense to know that someone else can look at the problem and possibly come up with a better solution.

Examples

The light plot shown in Figure 5-7 illustrates a typical rock band setup. Replace the keyboards with a steel guitar and you have a country band. This could even be a jazz group. The back truss and side light tree design is the basis for all concert design. Heavy back light is the mark of concert lighting.

The band back light washes could be split so that the drummer is separated, since he has no back light specials. Some people split the side light left and right so they can have, for instance, a red side light from stage left and an amber from stage right. The real controlling factor is the number of dimmers.

A more ideal control of these lamps would be:

> 4 band back light washes (4 lamps each)
> 4 drummer specials (1 lamp each)
> 6 band specials (1 lamp each)
> 6 lead singer back light specials (1 lamp each)
> 8 band side light washes (2 lamps each)
> 8 lead singer side light washes (2 lamps each)
>
> 36 channels of control

That quantity hits the nail on the head. Most consoles come in 24, 36, 48, 60, 72, or 96 channels of control. However, I insist on spare dimmers and control channels on the road. That way you can adjust at the last minute if a failure or other loss of dimmer or control should occur. Also, I make sure I have firmly fixed in my mind, well in advance of the tour, what control I could give up in order to solve the problem. Preparation is paramount. Using this plot, first I would give up side light circuits, second I would add drummer back light to band washes. In either case, the dimmer capacity must be high enough to accept this additional load, and planning can cover that point. Always have a few dimmers that can accept additional loads.

The colors I select change depending on whether the lead singer is male or female and even whether they are black, brown, or white.

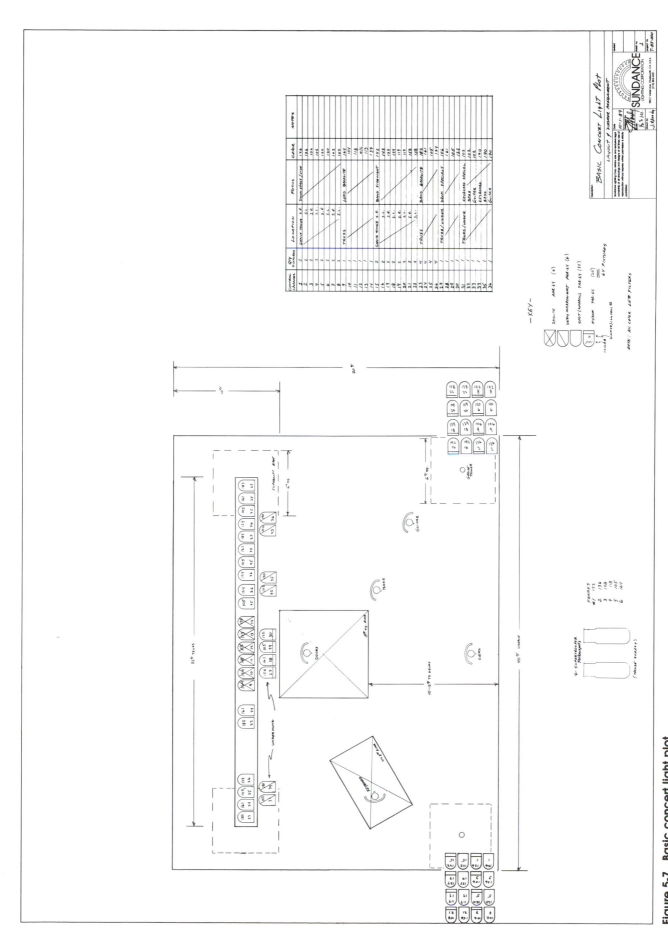

Figure 5-7 Basic concert light plot
(Designed by James L. Moody.)

Without going into the basic theory of color as it applies to theatrical lighting, it should be easily understood that any color (actually the absence of a portion of the visible light spectrum) projected onto a color will affect how the human eye perceives that color. Therefore, the designer must be sensitive to skin tone to most effectively illuminate the artist. The following color chart would be an acceptable beginning for most acts. Note that I have not designated a specific color media. At this point we are more concerned with the broader picture; exact color numbers are not yet important.

> Band & drum back light—red
> —blue
> —lemon
> —green
> Band back light specials—lt. pink or lemon
> —lavender or sunrise pink
> Band side light—amber
> —blue
> —blue/green
> —lavender
> Lead singer back light—red
> —lemon
> —lt. blue
> —lavender
> —magenta
> —no color
> Lead singer side light—red
> —amber
> —blue
> —no color
> —lavender

Color Changers and Effects

A few color changers (the scrolling type give six or more colors) can make even the simple plot in Figure 5-7 a very flexible design for a more advanced designer. (Color changers will be discussed further in Chapter 13.) The possible combinations that this multiplicity of color brings to the design would be much more advantageous on the road than the increased quantity of fixtures and dimming that would be required to equal the looks they can create. Remember that shipping space is critical.

Moving lights and truss-mounted followspots could also add to the flexibility of the system in Figure 5-7 without increasing the physical truss configuration. The pizzazz that is possible from even a simple layout can be used over and over again. Add the "toys" only after you have mastered the straightforward plot or you could get in over your head. Complexity can harm you more quickly than anything else.

Variables

Lighting fascinates me because it never looks the same twice, even when you are using a computer. The atmosphere of the room changes from hall to hall and your perception is slightly altered. Smoke in a nightclub or at concerts makes the light stand out more. Light from exit signs or candles on the tables changes the darkness level (black level, in TV parlance) from show to show.

Other variables are voltage to the lamps and changes in position and elevation. When you move the system from one hall to another, the voltage changes. Not many halls have exactly 120 volts at the input service—it varies between 108 and 123—and then when dimmers are put in line they will drop another 2 to 4 volts on the output side. Now you string 50 to 100 feet of cable to the lamp and it is rare to have 120 volts at the lamp filament. Therefore, the color temperature changes each day and the color won't be exactly as planned. With many gels, this will mean visibly altered colors.

Moving the trusses or truss upstage or downstage a couple of feet makes a difference in the angle the light will strike the performers; this alters the shading and contrast of your lighting. Another variable is in the followspot positions from hall to hall. Are they straight-on or at a 45-degree angle to the artist? Every variable will make a difference from show to show and must be considered.

Twelve people can take the example in Figure 5-7 and make 100 looks, but no two designs will be exactly alike. That excites me, and I think concert lighting releases the same creative imagination as theatre, because we must take 50 artists that all perform with similar staging and make them look different.

One final word of advice: experiment, experiment, experiment. I do, even with a show that is already touring. If I am running the board, I try things. Sometimes, it is better, sometimes it is a tossup, and often it is worse than my first idea. But I still try and I will keep on trying to bring clarity and definition to my design.

More Complex Designs

After you become proficient at a simple, straightforward plot, you can press on to some grandiose designs. The field of design loves the outrageous.

The freedom in creativity starts with the structures you design. Trusses come in all lengths, shapes, and load capacities. At this point, simply think of geometric patterns and designs (see Figure 5-8). These examples are only a few basic ideas for possible grids. The object is not to just make a pretty design as much as it is to choose light angles that create mood and drama.

While there is the practical limitation of load capacity, I can say that I have seen trusses that defy structural logic. In Chapter 11 you will be given the technical information you need to decide on a final truss layout.

However, at this "pure design" phase of our discussion, you can see how the structures become a real element of the design. Keep in mind that in almost all of the touring systems, the structures will be in view of the audience. So the layout of the structures can be both functional and an aesthetic element of your show.

Figure 5-8 Truss layout examples

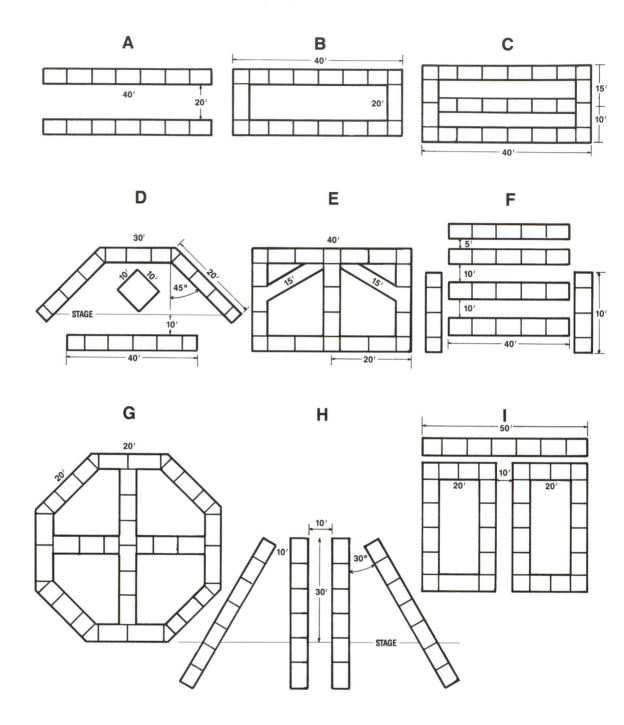

6

Cuing the Music

While doing the layout and color chart, you must be thinking ahead to how the cues will work with your plan. Does the layout give you the degree of flexibility you want? Does the color chart work as a palette in harmony with the music? A separate design criteria now enters the picture. The designer must take the raw data from the mechanical drawing and make it come to life in the same way a musician does with notes in the score.

I know that there are things that we do unconsciously. Someone asks us how we learned to do something and our answer is, "I didn't realize I had done that," or "I don't know how I did that." The fact is we did do it and we somehow learned to do it somewhere, sometime. One of the unconscious things good concert designers seem to do well is analyze music.

From the first time you hear the artist's music, you should be mentally planning the choreography of your lighting, breaking down the music into cue points. The analysis will help determine the type of lighting console you will require. The manual console is used by most tours. But in the case of John Denver, there were over 250 cues in the show. I determined early on that a computer was correct for the job because I wanted smooth, subtle cues that were repeatable. The computer allowed me to design each look and write my cues directly into the memory.

My years of experience allow me to take a few short cuts now, but I am sure I still take these steps subconsciously:

1. Listen to the song; try to pick up a lyric, a musical phrase, or the dynamics that make the general statement of the song.
2. Translate the song into a primary color.
3. Find the high point of the song (it may not be the end).
4. Find the repeating portions of the music, such as the choruses and verses.

This process has probably created four or five looks: the opening, the chorus, the verse, the solo spot, and the end. Changes between verse and chorus may be repeated several times. Also, a cue may occur at the *turnaround*, a musical device found in most pop music that allows the songwriter to repeat a melody by interjecting another musical phrase between the similar themes. Are the cues going to be bumps or slow fades? It does not matter what they are. What is important is that they each act as musical punctuation not just flashing lights!

After doing this for all songs, go back and look at the song order, or *set* as it is often called in music. See if you have the same color patterns for songs that are back-to-back. I do not hesitate to change the colors, to ensure that there is no repeated look in two adjacent songs. Certainly, a look can be repeated later in the set where it stands alone. If the color simply must be used, try to change the position that the color comes from, for example, when amber back light is used in the first song but the next song uses amber side light or an amber followspot and uses white as a back light.

How do you remember where the cues go? Obviously, they are not scripted like a play or an opera score. Some lighting designers, especially the Las Vegas designers, use lyric sheets (see Figure 6-1) much like a play is cued by the stage manager. This is a theatrical form rarely

Figure 6-1 Lyric cue sheet sample
A method for notating lighting, fly, and other cues when lyrics to the songs are available.

ALL NIGHT LONG (ALL NIGHT)

Cue		Part	Lyric
LQ 19 ↑(4)		BAND	(DRUMS 5 BAR INTRO)
LQ 20 X (2)		ALL	Da da OH
LQ 21 X (4)	FS# 1&2 ↑(2) F6 chest PU & ANDY	ANDY	WELL, MY FRIENDS, THE TIME HAS COME RAISE THE ROOF AND HAVE SOME FUN THROW AWAY THE WORK TO BE DONE LET THE MUSIC PLAY ON
LQ 22 X (1)		ALL	PLAY ON, PLAY ON
LQ 23 X (3)		ANDY	EVERYBODY SING, EVERYBODY DANCE
LQ 24 X (2)			LOSE YOURSELF IN WILD ROMANCE
		ALL	WE'RE GOING TO PARTY, KARAMU
LQ 25 X (2)			FIESTA, FOREVER
		ANDY	COME ON AND SING A LONG
		ALL	WE'RE GOING TO PARTY, KARAMU FIESTA FOREVER
		ANDY	COME ON AND SING ALONG
LQ 26 B↑	FS# 3&4 B↑ F2 Full FS# 1&2 X F1 pull Full ANDY	ALL	ALL NIGHT LONG (ALL NIGHT) ALL NIGHT ALL NIGHT LONG (ALL NIGHT) ALL NIGHT ALL NIGHT LONG (ALL NIGHT)
		ANDY	ONCE YOU GET STARTED YOU CAN'T SIT DOWN COME JOIN THE FUN, IT'S A MERRY-GO-ROUND EVERYONE'S DANCING THEIR TROUBLES AWAY
LQ 27 (2)	FS# 1&2 DO2 ½↓(2)		COME JOIN OUR PARTY, SEE HOW WE PLAY!
LQ 28 B↑ w/ chase "A"	FS# 1&2 RESTORE(3)	BAND ALL	(4 BARS) TOMBOLI DE SAY DE MOI YA YEAH, JAMBO JUMBO WAY TO PARTI O WE GOIN' OH, JAMBALI TOM BO LI DE SAY DE MOI YA YEAH, JAMBO JUMBO
LQ 29 (3) fade chase "A" (2)↓			OH
		ANDY	YES, WE'RE GONNA HAVE A PARTY
LQ 30 X B FQ 4 pipe #28 ↓(3)		ALL	ALL NIGHT LONG (ALL NIGHT) ALL NIGHT ALL NIGHT LONG (ALL NIGHT) ALL NIGHT ALL NIGHT LONG, ALL NIGHT , ALL NIGHT LONG
Pyro Q 3			(ALL NIGHT) (ALL NIGHT)
FQ 5 pipe #32 ↓(2)		ANDY	EVERYONE YOU MEET

used in concert touring but sometimes handy in longer running shows. Although it is not general practice in Las Vegas for the lighting director to leave after a day or two, the board operator could follow the lyrics and call the followspots and execute the board cues from those notations. I have done this on two occasions where shows were on long, open-ended runs and it worked.

One-night touring is different. Usually the lighting designer runs the console, but if not, the person doing so is traveling with the show and is as familiar with the cues as is the designer. If the designer is with the show and should know the music inside out, why bother with a formal cuing method? There are two good reasons: the possibility of illness or accident; and the fact that the designer is most likely working for several clients, and without total recall, it is hard to remember all the cues for each show. Also, after a few years of touring with an artist the designer builds up an extensive repertoire of songs not in the current show. As the tour goes on old songs are substituted or new songs are tried out. That takes a lot of mental agility to be on top of all that music. John Denver had over 50 songs the band could play and he often substituted without warning.

Forms

I use the 4 × 5 index card as a form (see Figure 6-2). Full-size 8½ × 11 pages are too big to place conveniently on the console and scan quickly. What kind of information you place on the card is the trick. I have a format I have used with a few variations for many years. There are two forms shown (Figures 6–3 and 6–4) and although there may not seem to be much difference, the detail in the preset column (column 4) is simpler for a computer, whereas a total breakdown of dimmer numbers and intensity is required for a manual board.

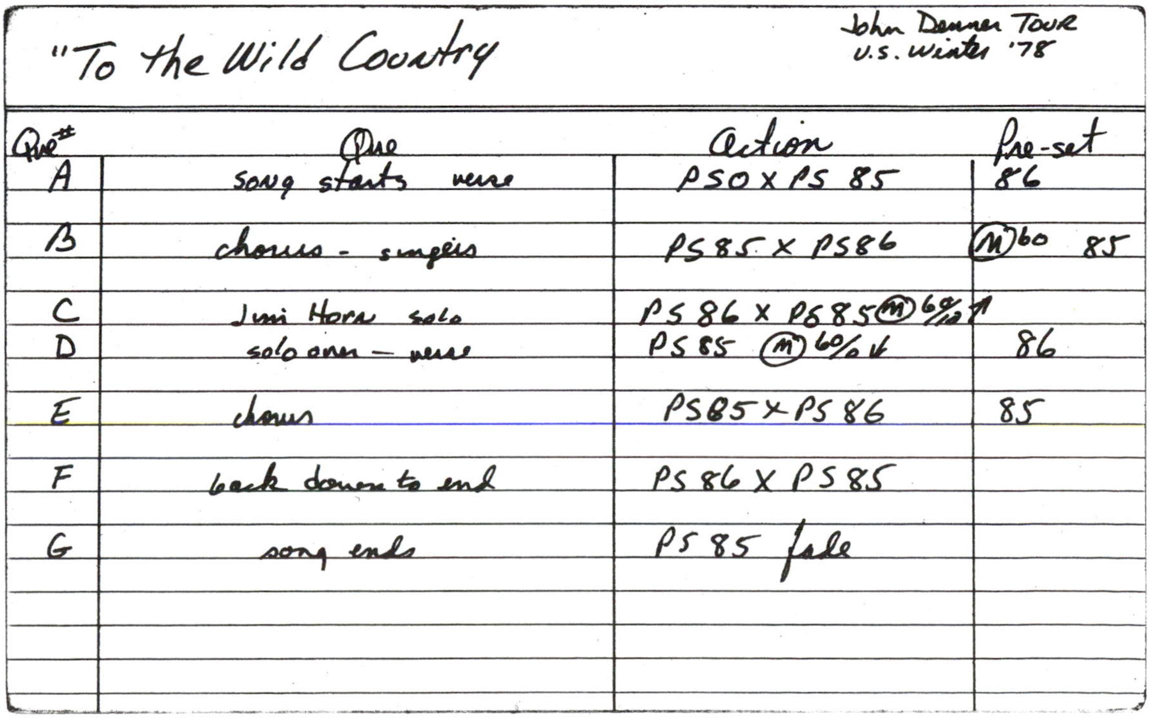
"To the Wild Country

John Denver TOUR
U.S. Winter '78

Que#	Que	Action	Pre-set
A	song starts verse	PS0 X PS 85	86
B	chorus - singers	PS85 X PS86	(M)60 85
C	Jim Horn solo	PS 86 X PS85 (M) 60/10 ↑	
D	solo over — verse	PS 85 (M) 60/10 ↓	86
E	chorus	PS85 X PS 86	85
F	back down to end	PS 86 X PS 85	
G	song ends	PS 85 fade	

Figure 6-2 Cue card, John Denver tour

Figure 6-3 Sample 4 × 5 cue card for manual board

"SONG TITLE"	MANUAL BOARD EXAMPLE		(SHOW ORDER) (Do In Pencil) (2)	
CUE	TIME	GO CUE	BOARD	F.S.
A	↑ (3)	MUSIC	4/6 10/F 12/8 14/7 20/F	1/F6 PU FULL
B	X (3)	VERSE	1/F 2/F 6/8 12/4↓ 22/F 23/8	
C	B↑	PIANO solo	27/F	B.O.
D	↑ (1)	VERSE	SAME AS B	RESTORE
E	X (2)	CHORUS	9/F 10/F 14/2 18/6 26/9	
F	B.O.	END		
G	↑ (3)	RESTORE		1/F N/C FULL

Figure 6-4 Sample 4 × 5 cue card for computer board

"SONG TITLE"	COMPUTER BOARD EXAMPLE		(Show Order) (4) (Do In Pencil)	
CUE#	TIME	GO CUE	NOTES	F.S.
10	X ↑ (6) (10)↓	SONG STARTS	WATCH DRUMMER	
10.1	ADD ↑ (4)	build band		
11	X (3)	VOCAL		1/F6 FULL 2/F3 CHASE
12	X (2)	chorus		1/F4 BX
13	X (4)	VERSE		X 1/F6 (4)
14	X (2)	chorus		
15	B X	guitar solo	STAGE RIGHT	1/BO 2/BO
15.1	B X	END solo		1/F6 PU guitar
15.2	B X	VOCAL		1/PU JOHN 2/F3 CHASE
16	X (2)	ADD BACKUP SINGERS		
17	X (4)	TAG	LEAD WITH F.S.'s	All FADE (3)
18	↓ (3)	END SONG		
19	B↑	RESTORE		

Notations

Cue Number

I use A-B-C instead of numbers for cues so that if I am calling board cues along with followspots, no one can confuse cue #1 with followspot #1 or frame #1. Also, I do not believe in consecutively numbering cues straight through the show, because the song order can and often does change during the course of the tour.

Cue

The cue is the downbeat, instrument, lyric, or whatever you use to indicate when the action should take place. Since the shows are not scripted, I use very simple ones such as "band starts," "first lyric," "sax solo," "chorus," "end song," "restore," et cetera.

Action

The action notations (under the time column in Figures 6–3 and 6–4) can change slightly depending on whether you are using a computer or a manual board. A computer board is simply noted, for example, P.S. 11 × 12 (2). That means preset 11 is to cross-fade to preset 12 on a two count (two seconds). A variation can be written: P.S. 10 ↑ (6)

and previous cue ↓ (10). That means to pile on preset 10 and fade previous preset on a ten count fade. Note: parentheses, ( ), always indicate time; the number inside indicates the seconds. A mark of B ↑ indicates a bump up, and B ↓ or BO are ways to indicate a blackout. Brackets, [ ], can be added for other notations. It is also possible on a computer board to be using a channel in the manual mode; that is indicated as M d36/7 (8), meaning: manually add dimmer 36 to a level of 70 percent on an eight count.

On a manual preset board, the notation could be A × B (3), meaning scene A cross-fades to scene B on a three count; or A ↑ B ↓ (3), meaning: add scene A and fade scene B on a three count. Another action would be A ↑ (3), meaning: add scene A on a three count.

Preset

The computer uses the column marked "Notes" (Figure 6-4) as a reference to the action or things to use to remind you of why the cue is done.

On the manual board, the "Board" column (Figure 6-3) indicates each dimmer level specifically, for example, d9/F d10/F d14/2 d18/6 d26/9 × (2), meaning: dimmer 9 level full, dimmer 10 level full, dimmer 14 at 20 percent, dimmer 18 to 60 percent and dimmer 26 to 90 percent, cross-fading on a two count. Similarly d 14/3 ↓ (10) would mean to fade out dimmer 14 on a ten count. This could also be used to lower the dimmer level, for example, d14/3 ↓ (5) from the previous level to a lower level. Since a fade out or change in level look the same in the notes, some people underline when it is a level change only.

Followspot

The notations under the followspot (FS) column can be complex if more than one reference must be considered. As a general example, however, followspot notations might read (john) F#6 ↑ (3), meaning: fade up on John in frame 6, full body on a three count. The use of B ↑ indicates bump up, and B ↓ or BO indicates a blackout the same way it does with dimmers. You can develop letter or numerical notes for the size of the circle of light needed; for example, I use "½" for a waist shot, and "HS" for head and shoulders. You can make up anything that is easy for you to remember when you are looking at your cards.

Miscellaneous

A miscellaneous column could be added for additional notations such as effects cues, set or curtain cues, warnings, et cetera. I do not use colored pencils on my cards, because it is time consuming and the colors are often misread under a red or blue work light. You may have been taught that "warnings" were one color pencil, "go's" another, et cetera. We do not normally have the script format to do that, so it doesn't work for me except in the Las Vegas venues.

Verbal Cuing

I have never seen a theatrical type cue light used on a concert. Intercoms have come a long way, and it is so much surer to have a response from the person on the other end. As far as I am concerned, cue lights would be worse than shouting for communication. I did one show where I had no two-way communication and I can only say that I felt as if I were utterly alone on the ocean calling out for help in the darkness.

The way you communicate to a houselight operator, a board operator, or a followspot operator will directly affect the show's smoothness and accuracy of cues. It is truly an art form unto itself. Chip Mounk has always been highly regarded for his effective followspot cuing, especially on the 1972 Rolling Stones tour. He is certainly one of the pioneers in concert lighting and developed many of the cuing techniques still being used.

When I was an air traffic controller, I learned something that is generally overlooked. Speech pattern, meter, and accent are very important. If the operator cannot understand you, he will not be able to do the cue properly. Your diction and local accent can frustrate the operators and reduce your comments to unintelligible noise. This is not to put down any regional accent; it is just a proven fact that how you speak will affect the understanding of the listener. The Air Force says that the Midwestern accent can be best understood by every English-speaking person. Meter is important because the speed and inflection you use in your speech will affect the operator emotionally. Obviously, if you are screaming or talking very fast over the intercom, you do the greatest harm possible, because the operator will tune out. He will think you are not in control—that is disastrous. As air traffic controllers, we were taught that sixty words a minute was the speed at which we were to talk; and I still find this a good speed for show cuing. I realize that at times you have a lot to say in a very short time between cues, but I would suggest that you take the time to find a way of saying it in fewer words, rather than speeding through to get it all in. After all, if you say it so quickly that no one understands you might as well not say it at all. All of these points can be worked on in a speech class if you have the opportunity, but you can learn by speaking into a tape recorder, giving cues, then playing the part of the operator. See if you would listen to yourself. Or have a friend who is not in the business listen and try to repeat what you say.

Cuing Followspots

I use the *key word* method of followspot cuing. This is a method of calling cues by using words such as "go" or "out" to prompt someone to react as planned, instead of relying on visual or mechanical signaling devices to call the cues.

My preshow speech to the operators goes like this:

Good evening ladies and gentlemen. I'll call the followspots as Spot 1, Spot 2, etc.

Indicate which light you are referring to; for example, northeast corner, left or right of stage, man in red shirt, whatever makes it obvious which spot is which.

I give all cues as: Ready . . . and . . . Go. Do nothing until you hear the word GO. It will indicate a color change, blackout, fade up, pickup . . . anything else you hear is for information or prepping an upcoming cue. All cue warnings are given as: standby Spot 1 in frame #6 waist on Fred for a 3-count fade-up . . . Ready . . . And . . . Go. The counts are in seconds and all fades should be evenly executed throughout the numbers.

Then verbally count off so they hear at what speed you count.

> The stage is set as follows . . .

Here I would lay out the placement of all players, including items that would help them to remember each musician.

Some people use theatrical letters for areas of the stage. If it is necessary because the artist moves around the stage a lot, do such a plan in four or five areas across the stage and two or three areas deep. To close, I say:

> Thank you ladies and gentlemen. I'm looking forward to a great show. Take your color and head for your lights; check in on the intercom when you are in position.

While I do not use operators' first names to call cues as many people do, you should do whatever is most comfortable. Using their names makes it more personal and eliminates one number that could be confused with all the other numbers being heard by the crew. During the show, I tend to talk a lot to the operators, mainly to give them a feeling of being a part of the show and not just treated as robots who turn on and off the light. And, if you find that the operators like to chatter a lot, keep up a running dialogue yourself so you can call a cue anytime it is needed. But be careful not to confuse them as to what is your chatter and what is actual cue information. I find that better than telling people to "shut up." Give people credit for their skills, and they generally respond with their best effort. If they have a general idea of what is about to happen, they relax and the tension is relieved. Certainly, this is not always the case, and I have been in positions where the less said the better; your own judgment must be used in each show to feel out the situation.

Summary

Whether you use cue cards or other forms, the important point is clarity and being frugal with cue words and notations. The cards were not designed so that another designer and board operator could walk in and take over right away. The cue cards are a personal preference; they are only there to assist me in running a smooth show. When I use board operators, I do not write out their cards. I use mine to talk them through and let them do their own cue notations as they choose.

In the final analysis, the smoothness your show has will be directly related to your ability to cue both board and followspots in time to the music. The aesthetics of your choice of colors and angles will affect the audience, but not nearly as memorably as a late or missed cue. The rock and roll audience has become very sophisticated and realizes that the lights are a very important part of the show.

Cue a show well and the artist will get better reviews than if you are lazy and off tempo. The psychology imparted in what we do cannot be underestimated.

7

Road Problems and Scheduling

Timing is the key to efficient production of a concert tour. Most problems during setup grow to become major disasters, not because of faulty equipment or damage that cannot be repaired, but because you only have fifteen minutes to an hour to fix it. If that is not bad enough, you are in a strange town, usually close to or after 5:00 P.M. Time is your biggest enemy. Effective handling of your setup time will give you more time to discover the problems and, when needed, to get parts and make repairs in time for the show.

Time is also the most important factor when a truck is late. A show normally loads in at noon and gets a sound check done by 7:00 P.M. (on average and depending on production complexity). When the truck does not get there until 4:00 P.M. or 5:00 P.M., can the show be ready? It had better be ready! Here is where your organizational skills and efficient use of local resources really shows. In general, problems on the road can be handled on a day-to-day basis if you keep calm and face each one with your mind open to several alternate solutions. Then make that decision quickly! The worst thing to do is to delay making a decision. There is no right or wrong—some decisions are better than others in retrospect. But in the end, the show will start at 8:00 P.M., hopefully with sound and at least some lighting. How you accomplish the tasks you are assigned makes you either a successful road designer or a failure.

Power Service

After timing problems, the second most important problem can be finding a good power service to use. This is especially true at colleges. Most schools do not have their gyms equipped with a bull switch or other power connection device for use by road companies. Normally, power is taken off the houselight panel. That is very bad electrical practice and, in most instances, a code violation. But when there is no other power within 400 feet of the stage you make do. This is an area that needs a discussion all its own. The best answer is to be as knowledgable as possible: read up, or better yet, talk to a licensed electrician who does temporary service connections for a living. Make it a practice to require a house electrician be on hand at load-in. Even though you probably do not need power for an hour or more, it gives you time to discover if there is a problem and then time to solve it without delaying your schedule. Also, if rigging is to be done, the crew will need power for their rigging motors as soon as they start so as not to delay the load-in.

Followspots

Followspot and operator problems seem to run with power problems. Generally, if the facility does not understand your power requirements, it won't have good followspots or operators either.

A good road technician can diagnose what is wrong with a followspot without being at the light, and is able to tell the operator how to correct the problem. Talking operators through the cues is another problem. Since the followspots are the one lighting element not generally carried with the show (except those mounted on the trusses), the efficiency of the unit is tied to the operator's ability to run it. Unlike the Broadway show where several days of rehearsals with followspot operators are accomplished before opening for previews, the concert starts with only a brief talk to the operators (see Chapter 6). When you give the first cue, you see simultaneously how the operator handles the light as well as the efficiency of the unit. There is no time to replace or repair the unit. I can safely say that followspots have caused me more problems than any other element of a show. They are the great unknown factor.

Stages and Ceiling Height

Stages themselves can be a major problem. Although size is often unknown before arrival, it is usually the way the stage is built that comes under attack. When the crew arrives to find an unsafe stage, either because it is not braced properly or because it is uneven, the delay in getting it corrected takes away valuable time. Size variations should be considered in advance so that the set and lights can be adapted—within reason. Large public arenas and auditoriums have sturdy portable stage structures. It is when the stage is built by people unfamiliar with the devices that are to be placed on it that problems can occur. Concert artists generally bring complete portable lighting systems and, if ground supported, these structures add a tremendous weight to the stage. I have actually been told that the house crew thought everything would be okay because the stage had been big enough and strong enough for the "Innercity Symphony."

Ceiling height can also be a problem. This usually happens in clubs and small facilities that were designed as multipurpose rooms. The ceiling may be twenty feet high, but put in a six-foot-high stage and you can only put lights up twelve to fourteen feet above the stage. You must also consider what the ceiling is made of; can it take the heat that the lighting fixtures give off? This low height is an enormous problem for back lighting, because the lamps will be at about a twenty-degree angle to vertical (normally it should be thirty-five to forty degrees). This will spill out into the audience a considerable distance, causing people in the front sections to have difficulty viewing the performer.

Special Effects

Projected effects or special effects such as smoke, fire, and explosions can cause problems. Projected effects usually require a bigger or deeper stage. What happens when the stage just cannot be forty feet deep, the screen cannot be moved upstage any further, or the projector cannot be rigged where it is supposed to be placed? The show should carry both rear and front screens for projection. (Actually, a show carrying these effects should have had sense enough to check out these problems in advance.)

Pyrotechnical effects are something else again. Most cities require special licenses or permits to have flash pots or open fire on stage.

Some cities flatly forbid the use of fire on stage. Where permitted, these effects must be handled with extreme caution. The best example was seen worldwide during the California Jam at Ontario Motor Speedway in the summer of 1974. The English group Deep Purple used flash pots in their act. The effect was to happen at the end of a guitar solo when the musician was to destroy the guitar and amplifier (fake) with the result being an explosion. After the licensed pyrotechnician had loaded the device with the charge as per regulations set up by the State of California, it was believed, but never proved, that someone with the group decided it was not powerful enough and added more powder to the charge without informing anyone. The result was seen by the quarter of a million fans at the concert, and by millions of people who saw it on the television special. The stage caught on fire; men with fire extinguishers could be seen running on stage to put it out and that was not in the act! If the concert had been indoors, the potential for disaster would have been much greater.

Road Life-Style

A lot has been written about rock stars and their drug use, sexual activity, and money spending. Because the technician spends sixteen to twenty hours a day with the artist, some of these excesses can rub off. The problems begin when the technicians and designers forget why they are really there. It is not to be buddies with the stars, nor to see how many pickups they can make at the concert. The good technician or sound engineer should consider the mounting of the show as the real reason for being there.

Drugs and booze are both problems. A drunken technician is just as likely to make a mistake and hurt someone as a drugged one. Personally, I have had more trouble with drunken crewmen than drugged roadies.

The actual workday can be shown best by giving a typical schedule. Consider that this schedule is repeated five to six times a week for six to twelve weeks at a stand to really get a feeling for the strain and hardship this puts on people.

Road Timetable

6:00 A.M.	Wake-up call
7:30 A.M.	Depart for airport
8:00 A.M.	Arrive airport; check bags and turn in rental car
8:30 A.M.	Flight departs
11:00 A.M.	Arrive next city
11:20 A.M.	Rent cars and get bags
12:00 noon	Arrive at hall (seldom time to check into hotel), begin setup
4:00 P.M.	Sound check (all lights, sound, and band equipment ready)
5:30 P.M.	Reset band equipment for opening act
6:00 P.M.	Opening act sound check
7:00 P.M.	Open house, crew meal
8:30 P.M.	Show starts
11:30 P.M.	Show ends

| 12:00 A.M. | Load-out begins |
| 2:00 A.M. | Load-out complete, go to hotel |

A person who goes out with a show like this must be able to handle himself or herself physically as well as mentally. The body can only take the pace for so long. You should prepare yourself as best you can, eat as regularly as possible, and organize your sleep/play time to benefit yourself the most. Sometimes that is play, because the body needs that, too, but most often it will be bed . . . alone. Living out of a suitcase can be much harder than the physical pace of the tour. I know many road technicians who do not have a permanent address. Most humans need a place to call home, or someone waiting for them to return. The road technician often has neither.

Even as I say these things, I know that most readers of this book will believe that the excitement of being part of this piece of our youth culture and theatrical history outweighs all these hardships. Well, maybe that is what keeps us going. The newness of each day is at the top of my list. I do not believe I could be happy going to the same facility each day. The job certainly is not for anyone who wants security and a nine-to-five job. The day starts and ends in work-related activity. Very few minutes are spent that are not tied to the group or the people who are traveling with the show. You had better all get along or tempers will eventually flare.

Transportation

The methods of transportation are changing. In the early days of touring, it was "in" to fly everywhere. But the economy has finally caught up with us. Rockers now see the value in what many country/ western artists have known for years.

Custom-built buses and motor coaches outfitted for sleeping and riding in comfort are more practical. The time taken in running for airplanes and the questionable safety factor of flight is now being replaced with the ease and security of the bus. After all, there has never been a rock star killed while riding on a bus!

Ninety percent of my tours since 1978 have used private coach or motor homes as the primary means of crew transportation. If the truck can make it with the equipment, so can the bus. So, we have returned to the old "bus and truck" touring companies of ten years ago. The difference is in the comfort and class of the bus interiors. These custom coaches are hardly Greyhounds. They are completely custom designed and fitted with sleeping sections as well as front and often rear lounges that have everything to keep the band and crew occupied on the trip. Normally they handle eight to twelve people comfortably for sleeping.

Figure 7-1 shows a Silver Eagle bus modified for touring a band or crew in comfort and safety. Only the shell, engine, and drive train come from the manufacturer; the custom-coach builder does the rest either to his own design or often to specifications provided by the artist who will use the bus. In Figure 7-2, the view looking toward the driver shows part of the lounge. Most are outfitted with microwaves, refrigerators, TV, VCR, and stereos. Aft of the photo would be the sleeping area. The rear can be a second lounge or office.

Although some artists can afford a private jet plane, few groups are in this financial position today. More and more are going to scheduled airlines or ground transport such as the custom coaches.

Figure 7-1　Custom motor coach, exterior
(Photo by Florida Coach, Inc.)

Figure 7-2　Custom motor coach, interior
(Photo by Florida Coach, Inc.)

An Established Field

Touring is now a well-established theatrical field. Within the past five years it has been shown to set entertainment standards that have heavily influenced television, film, and theatre. Touring has proven to have lasting entertainment and cultural value. The economy will always have its ups and downs, which will affect what the ticket buyer can afford to spend on entertainment, but it has already been proven that concerts hold their own, even in slow economic times.

The artist who is now a superstar will continue to draw huge crowds, and, it is hoped, will continue to increase the production values—and budgets. The artists who fall into the production/show group will have to spend even more production dollars to draw crowds. The 1987 David Bowie tour cost a reported one million dollars a week to put on the road. Financially, new bands are the ones that will be worst hit by increased production costs.

But whatever the state of the economy, the production values, budget, and complexity of design continue to expand in the concert field. And with this expansion will be a crossover in the use of these techniques in other media. The technician or designer who sees the value in the advances made in concert lighting will be better equipped to deal with problem solving in other theatrical forms.

8

Road Safety

Anyone who has been on a stage realizes very quickly how dangerous a place it can be if you do not watch your step. Scenery moving, pipes flying in and out, trap doors opening, or risers that are not stable: all are accidents waiting to happen. So many incidents are reported that many states have tried to enact hard hat area laws. Even venerable Broadway gets its share of accidents and deaths each year. Stars are not immune—I remember when Ann-Margret fell from a platform at Caesar's Palace in Las Vegas while rehearsing. In that case, a quick-thinking stagehand broke her fall and possibly saved her career, if not her life. He was injured for his efforts, as are many stagehands yearly.

It is hard to find a stagehand who has not had a personal mishap or injury or does not know of a friend with one. Usually they amount to minor injuries, but it points up how often they occur.

A sizable portion of these injuries happen during set changes on a darkened stage. When you only have seconds to make a scenery move and one person is off the mark that night or there is a substitute who has not done it before, there is the potential for error and possible injury. Working with local crews who have not seen the show before and are doing in one day what theatrical shows are allowed to rehearse over and over cannot help but be trouble. That more accidents do not occur is reason for kudos to the road crews and local stagehands.

Even so, the road offers enormously greater potential for accidents than does general theatrical production. When you consider the long hours and the travel, fatigue is a big factor. The body just cannot stay at peak performance week after week for months at a time. The pressure on the traveling road crew is tremendous. On the average, they must get the show loaded in, rigged, and sound checked, then through a performance, strike, and load-out in under twelve hours and repeat that process five to six times a week for six to eight weeks at a stretch.

One of the major concerns of the now defunct Professional Entertainment Production Society (PEPS) when I helped to found it in 1980, and subsequently served as its first President, was the safety of road crews.

Not only was our own personal safety of concern, but we wanted to promote the image of people who wanted to work as safely as possible. We felt that as a whole, our fellow road technicians had an outstanding safety record. But there were no statistics to support or refute our beliefs. Compiling such figures is not easy for a small, new organization, because even if the membership were polled, it would not be a large enough percentage of the total industry. The cost of a properly supervised survey was beyond our means. Nevertheless, our informal surveys seem to back my original belief.

Our concern for safety was so great that as a primary requirement for membership a company had to have a $1 million public liability insurance policy. We wanted the hall managers, promoters, and producers to see our concern for not only our own safety but for the audience's well-being.

A safe working area is critical. Our area is the stage. But what goes into making it safe? Personally, I find that most minor accidents happen because people just did not think. We cut ourselves or bruise a leg and then say about ourselves, "What a dumb move," and that is exactly right. That implies we really knew better than to do what caused the injury; so why do we let down our guard?

A large percentage of the incidents can be attributed to doing things too fast. We are obsessed with how quickly we can get a touring show set up or loaded out. That haste can be a major factor in an accident.

The constant long hours of show after show in different cities without a break can not only be physically tiring, but mentally boring. You get into a routine and do things mechanically without thinking. There is a positive side to that, but it could also be fatal. The positive side is that with so much to do and so little time in which to do it, things must flow and fit together without a lot of effort. In most cases we do not even count items in a case; we "just know" it is all there. Instinct, I guess, but it also comes from repeating the process so often that it becomes an unconscious action. This is not meant to encourage beginners to disregard checklists and planned procedures, rather to point out that disregarding these could be very dangerous.

Specific Areas for Safety

Where are the key areas to watch for potential accidents? The answer is: anywhere they are not expected. However, there are some specific areas worth discussing, as follows.

Truck Unloading

People who are brought in to load and unload trucks have no idea of the weight of an item, what is in a box, or how delicate the contents are; they are just trying to get the truck unloaded. A forklift driver cannot be excused for dropping a case, but the loader who is trying to get a top-heavy case down a narrow ramp is always in danger of its tipping over. Make it a practice to have one member of the road crew outside the truck to assist the loaders with information about case contents, weights, et cetera, and to direct the movement of the cases into the hall for an organized placement. Another crew member should be inside with the loaders.

Rigging

When overhead rigging is required for the show, most often a professional theatrical rigger is traveling with the show. The rigger may also have a second or *ground man* with him. That is the person who sends up the cable, bridles, shackles, chain, et cetera needed by the rigger or *high man*. I feel this person is worth the added expense, especially during load-outs. The correct packing of the cases can save hours on the next load-in. Simply by eliminating the wasted time and energy of locating parts that were not packed in their proper place makes for a quicker, smoother, and safer operation.

If the show does not travel with a qualified rigger, it is wise to have one go over the plans before the tour starts and suggest a rigging plot, or, better yet, draw one out, giving insights into point placement, position, and weight configuration. This greatly helps the local rigger by providing a better idea of what must be lifted, thus promoting knowledgeable decisions based on load limitations at the specific facility.

After the rig is flown, always have the crew place several *safeties*. These are non-load-bearing cables attached between the portable truss and the physical building structure. They are there in case a chain or cable slips or breaks. A few years ago a truss did break as a safety was being released after a show. If the safety had not been there, it could have happened while the show was in progress, causing injury to artists as well as audience members.

Stage Support

When ground support for a truss is used, there is often little possibility of safeties, but you can still maximize the safety of the system. Check the stage surface carefully. When on a portable stage, especially check for weak spots or uneven sections. Next look under the stage. Are the sections secured together? Is the deck secured to the legging? You do not have to be a structural engineer—just common sense will tell you if it looks unsafe. If so, tell the promoter or facilities staff about it and make them change it. Even though it is not a direct concern of the lighting technician, I always slide my foot over cracks and joints to check for leveling and open spaces that someone might trip over. A portable "dance floor" material is a big help in solving this problem quickly and at the same time gives the stage floors a better appearance.

Ground Support

The types of structures that are commonly used to support lighting are shown in Chapter 12. Let us look at the safety involved in their use. Some of the structures have safety devices built in, such as with the Genie Super-Lift. A breaking device is designed to prevent the load forks from falling in the event the cable breaks. That lift is probably the best unit employing a cable and winch design with regard to safety currently available.

The hydraulic lifts do not have that particular safety problem, but they have greater problems related to leveling. They are so heavy that they can tip over if not properly stabilized or, if there is a weak point in the stage, they could break through the floor. In that case the lift did not fail, the stage did. Nevertheless, it is a safety problem related to the device and the stage. Use outriggers or stabilizer bars on all lifts to get the widest possible base.

Probably the two most common accidents that occur with lifts are: pinched fingers (getting them between columns or joints as the device comes down faster than expected); and shin bruises when someone runs into a lift's outrigger leg.

Lastly, no matter which lift is used, recheck leveling once the truss or load is raised to ensure that it is level.

Fixtures

Safety devices for fixtures are not common. Most PAR-64 fixtures have a spring-loaded clip to hold in the gel frame. But I like to take a tip from television and film lighting, where it is an accepted practice to safety chain any barndoor or snoot to the fixture's yoke. Usually, plumber's chain and an S-hook are used. I prefer wire rope with a ring attached to a corner of the gel frame and a dog-clip on the other end. If color changers or other effects are added to the fixture, a similar device should be used.

When working with a triangle truss, I also safety all fixture yokes to the truss or pipe. If you are using lamp bars of four to six fixtures attached to a uni-strut track, a safety should be added between the bar and the truss.

Another major area of danger is from lamp failure. The lamp manufacturers refer to this as *violent failure*, not an explosion. It can happen with an ellipsoidal or Fresnel lamp, but because of the lamp housing design, it would be rare for glass to fall out of these fixtures. However, the PAR-64 lamp has little or no real protection for this eventuality. The most extreme measure is to place a wire screen in the gel holder. Other than this, there is little that can be done. Hot glass showering down can cause serious burns and cuts.

The problem is intensified by the fact that quartz lamps are sealed under pressure. Therefore, a crack in the quartz glass acts just like puncturing a balloon—it explodes! The PAR-64 and the other lamps in this family have a glass lens sealed onto the face of the reflector-coated glass backing. Also, if the cement has not completely sealed the gap, the heat created by the lamp can expand the glass, allowing pressure to escape and causing the further fracturing of the seal, which could lead to an explosion.

Now, this is a layman's view and should not be considered a technical evaluation by an expert in lamp design, but it covers the popular views. The lamp manufacturers have taken great pains to ensure that the lamps are airtight before shipment. However, bouncing around in a truck can cause damage or cracks to occur. I have seen lamps come out of a truck neatly separated at this seal, and the lens lying against the gel frame. It is not practical to check each lamp every time they are moved because visual inspection will probably not reveal any fractures. This is one area of safety concern that continually plays the odds.

The PAR-64 family has another problem in that it is focused by manually rotating the porcelain connector cap at the back of the fixture. These caps are constructed of two pieces of porcelain held together with nuts and bolts. In transport they can come apart, thereby exposing the metal contacts. Without looking inside, people reach in to rotate the lamp and can come in contact with these electrically hot leads. Always look before reaching inside or at least wear gloves as an insulator. It is wise to use gloves in any case, because the lamp housing is very hot.

Another part of your daily inspection of the truss should be for socket caps that are off the lamps or that have come apart. One method used to keep these caps from separating is to use a high-temperature glue in the hole that contains the hex nut. I have also seen wire wraps tied around the two parts. There could be a heat problem with this method and potential fire, so I do not recommend it. Recently, some manufacturers have begun to install caps that go over the porcelain. It is a great idea and should be retrofitted to all PAR-64's.

Focusing

This is the most dangerous time of the day. When someone is high on a ladder or walking a truss, wrenches can slip out of hands, gel frames can fall while changing color, et cetera, so use extreme caution on stage. Warn people to keep clear. The falling wrench problem can be minimized if you attach a line to the handle and around your wrist. On tour, it is extremely hard to keep people out from under the working area, because so much is going on about the stage at the same time. When something falls, generally the person hurt is the stagehand holding the ladder, being directly underneath. He or she should have been looking up anyway, but that is not an excuse for the accident.

A constant problem is getting ladder movers (holders) to pay attention. If you are the person on the ladder, always make clear, before you climb up, how you will direct the movements. No one else should call for the ladder to be moved, only the person up high. Try to get at least one person to keep looking up, both to hear you better when you call an order and to watch for falling objects. If no ladder is used, and a person is up high on the truss, I always station a stagehand to watch for falling objects and to keep people away from the area.

Trusses

A daily visual inspection of welds is a must! A fracture can occur in the best of units but severe road conditions can do great structural damage. I believe that X rays of trusses should be a yearly requirement. This would reveal any hairline fractures before stresses cause a serious structural failure.

Second, and no less important, is to have one person responsible for rechecking all connectors—nuts and bolts, pins, and so on—that join truss sections together before raising the truss in the air.

Power Hookup

The potential for an accident with power hookups is tremendous. Since we are dealing with temporary hookups, it is not always possible to be sure the correct connection has been made. Another problem is that many situations require that the road electrician give the pigtails to a house electrician to do the actual hookup. Take the time to clearly show the color code markings on your pigtails. Do not assume everyone knows green is ground (or earth). What kind of ground is available: equipment ground, grounding rod, or water pipe? Is the service single-phase or three-phase power? Check all of these things carefully. Then recheck everything at the source after the hookup is complete before energizing the lighting system. I strongly recommend volt meters on each leg of the portable dimmer system you are taking on the road to double-check power before the system is energized. At least carry a good quality volt/ohm meter.

Grounding is the number one cause of power problems. The two main problems are potential shock to someone working or touching the lamps, truss, or lifts, and a buzz in the sound system created by problems between lighting and sound grounds. The necessity for a good ground is essential, especially outdoors. Many road electricians take the added precaution of grounding the lifts via a ground stake at outdoor shows. It is an excellent practice.

Murphy's Law

Murphy's Law—"If anything can go wrong, it will!"—is certainly true on the road. We place terrible strain on equipment and personnel. Fatigue can bring Murphy's Law into action without warning. The biggest deterrent to accidents is to be aware at all times that they occur when least expected. So expect it or you may be a statistic. This is why I place so much importance on stress and fatigue on the road. You may have done the setup a thousand times and can do it in your sleep, but if you are "asleep," you are an accident waiting for a place to happen. Insist that everyone on the crew be alert, because someone else's carelessness can be the reason you get hurt. I have at times told people, "You might not care if you hurt yourself, but I want to be around to do this job for a long time."

Safety Problem Corrections and Solutions

There are four areas of concern in maintaining safety. First, while still in pre-production, check the certification for load capacity of the truss design, and check the safety record of the company or equipment to be used. Has the equipment been inspected recently? Even if you are "only the designer," you do have a moral as well as legal obligation to ensure that what is used on the production is safe. Do not try to put this responsibility on the head electrician. Everyone must be concerned with safety. As the designer, you must know that the design you have submitted can be safely executed by the equipment supplier.

Second, as the designer or crew chief, you have a continuing obligation once on the road to make sure the equipment is maintained and the crew is working in a safe manner. The third area relates to the safety of the work environment: specifically the stage upon which the touring production is to be performed. Even if there is no lighting placed on the stage, as you walk around you should be conscious of wobbly sections or other signs of an unsafe stage. Report it to the stage manager.

Finally, make sure the equipment supplier has proper insurance on the equipment as well as for the people working for them. Legally this is an area that is not clear. If you come into a facility and see or suspect an unsafe condition and report it, you cannot relinquish your responsibility if the house manager, promoter, or stage manager tells you it is okay or to mind your own business. If the safety problem is in your area of expertise, you could be held criminally negligent if an accident happens subsequent to your making others aware of it.

If you are going to be an independent designer, I suggest very strongly that you obtain a public liability insurance policy. You owe it to the people you are working with and to the public your work comes in contact with to protect them.

I do not know of anyone ever being sued over this point, but legal counsel for PEPS had advised its members of their ultimate obligation in this matter. You are responsible for the safety of others. Insurance companies pay big settlements on accidents arising from people tripping on cables or having things fall on them during shows. Do not put yourself in the middle; avoid such a possibility by making your production as safe as possible.

9

Dealing with Problems

The human brain has selective memory; we tend to forget, or at least push into the far recesses of our minds, the disasters we've all faced. You know the type: the important lamp that blows just as the curtain goes up, or the color that looked great in rehearsal but now seems washed out. These things happen to every lighting designer and are unavoidable. The important thing for a designer or technician to know is not how to place the blame or make an excuse, but how to deal with the problem as quickly and effectively as possible.

Problem Solving, Stress Management, and Interpersonal Communication

The best designers are skilled in the art of problem solving. They communicate well and possess the ability to make on-the-spot decisions. They do not procrastinate. I've found that, on a day-to-day basis, I choose to work with other designers who possess communicative and decision-making abilities over more creative persons who cannot verbalize their thoughts and ideas. Concerts are a collaborative venture; when one member of a creative team is unable to deal effectively with the others, that person drains everyone's energies.

Let me tell you a story. While I was in graduate school I was assigned to stage manage a show. The lighting designer assigned to the show was very highly regarded by his professors, who thought he had a bright future.

On this particular production, the director was a visiting professional. He took a nontraditional approach to the play, forcing us to look at it in a new way. In the early meetings, we all heard his explanation, and in subsequent creative conferences, the design team contributed ideas. When it came time for the lighting designer to explain his thoughts, the director listened carefully. After hearing him out, he commented on his approach, saying everything was fine except for one key scene, which he asked the designer to rethink. In the next meeting, the lighting designer reiterated the exact same concept. Again, the director said no, rethink it.

Technical rehearsals of the play began, and guess what? The look the director had rejected showed up on stage. Again, the director was patient—an uncharacteristic quality among his peers—but said, "Okay. I've seen it, and it still does not fit the imagery I wish this scene to create. Show me something different tomorrow."

As stage manager, I got a crew call together for tech cleanup the next day before the dress rehearsal. The electrical crew was on time—but no lighting designer. We waited two hours, and still he did not appear. The board operator said, "What do we do? The director will hate it if I bring up the same preset again tonight." Figuring the designer had an emergency, we decided to cover for him and programmed a different look before the rehearsal began. At rehearsal time, the designer walked in. No emergency; he simply said, "I went to the beach to think."

We need not carry this story any further, except to say that this highly promising designer never made it past small community theatre. His talent was wasted, to be sure, but even more tragic was that his teachers did not educate him in the art and science of problem solving, stress management, and interpersonal communication.

Decision Making

As I mentioned earlier, part of my life was spent as an air traffic controller. How that happened was simple: I joined the Air Force, and they said, "You're an air traffic controller." Life often plays tricks on us; we may see no reason for certain events as they're happening, but they turn out to be pivotal influences later on. That was the case here, because air traffic control is essentially a game of juggling schedules and making commitments. The greatest lesson was:

Do Not be Afraid to Make Decisions

Even the smallest hesitation in committing yourself, based on your training and experience, could be fatal to the people on the plane. Therefore, a lot of time was spent *teaching* us to make decisions. The ability to make decisions is a learned skill. And, working under great pressure is equally critical.

The failure of our educational system is the lack of instruction in the psychology of action or decision making. Theatre, film, and television are action-oriented professions. We all must work to a production timetable that, it is hoped, brings all the technical and acting elements together at the same time. My earlier story is a classic case of a creative mind unable to deal with the realities of group-created art.

The best designers seem to be those who work well under pressure. Sadly, many who have the creative and even the communicative skills necessary to be good lighting designers cannot deal with pressure. The best lighting designers know how to handle disasters, both physically and psychologically.

What do you do when you encounter a problem—say, discovering you've used the wrong color? The solution is simple: admit your mistake and change the color. For most people, the problem is not in changing the color, but in admitting they could have made a better choice. It is no different when an actor cannot seem to find a special. Though it may be the actor's fault, I do not even think about arguing the point with the director. Change the special. And don't think the crew is standing behind you laughing; they'll only laugh if you are too foolish not to change it!

Design, Crew, and Equipment Failure

An important thing to realize is that we fail as often as we succeed. It's not all black and white either, because every success contains elements of failure, and vice versa. The types of failure lighting designers commonly experience can be put into three categories: design, crew, and equipment.

Design Failure

There is no such thing as a perfect design. By that premise, all of our works fail in some way. Teachers, critics, audiences, producers, and directors all decide in their own terms whether a production succeeds or not, and their evaluations are important, especially in the professional world. But honest self-evaluation will do you the most good. Learn to step back and go through your own checklist, asking if the show worked for you: as the director or artist defined the problem, as the physical limitations of the production required, and as you conceived it on paper.

Ask yourself what you learned from this design experience. What could you do differently next time? Each production requires you to rethink ideas or solutions that may have worked perfectly in other situations. To think you will succeed on the basis of your past glories will be the death of you as a creative person.

Crew and Equipment Failure

Regarding the crew, ask yourself if they failed, or if you failed because your training, your supervision, or your communication skills were inadequate. When crew members fail, look to yourself first and then *do not look any further!* Did you give them all they needed to succeed? If they failed, it was probably because you did not communicate your needs adequately. You would be wise to study the many good books available about motivating coworkers; I particularly recommend *People Skills* by Robert Bolton, published by Prentice-Hall.

I do a lot of one-day video shoots, and often encounter problems that would be easy to blame on the crew if I chose to do so. It's not uncommon to be short an adapter or cable, or to forget a lamp, especially when you're working on a limited budget. Obviously, you can't bring the whole warehouse. Most likely, the problem is that I've put more effort into the complicated production and did not spend a lot of time working out the nuts and bolts of this "simple" shoot.

Why waste time fixing blame? Do something! Maybe you can gang circuits, or look for a place where a fixture can be moved and used to solve the problem. Maybe you did give the correct list to your gaffer or the rental shop, but that does not help you now. Before the next show, say a word to the person who you think forgot to check the equipment. He'll appreciate that you did not jump on him in front of his coworkers and will be more careful the next time.

No one is necessarily to blame for equipment failure, but you are responsible for finding a quick solution. While doing a series for the USA Cable Network we shot more than 120 half-hour wraparounds. Those are the in-studio segments that lead into and out of a field report such as you see on *60 Minutes* or *20/20*. These had been shot in groups of twenty with a month off in-between. When we came back for a fourth session, I put the lighting in and couldn't get the same intensity on one part of the set dressings, even though we had the same studio, same dimmers, same fixtures, and same set. Why? We never figured it out. Under pressure to get tape rolling, my solution was to add a small fixture that was available and, to get the intensity, I put it at spot focus. About an hour into the taping, there was a violent failure (see Chapter 8). Luckily, the Fresnel lens kept the hot filament fragments inside the instrument. We quickly changed the lamp and went on taping. An hour later, it happened again. We then exchanged fixtures, and the problem was eliminated.

When the producer asked, "What the hell is going on?" I could have blamed the lamp, the fixture, or the equipment rental house. What I said was, "I screwed up." Finding a scapegoat or going into a long explanation on focus and lamp failure would only have extended the problem and wasted everyone's time.

Dimmer Problems   Here is a common problem: dimmers that develop minds of their own. It doesn't happen often in permanent installations, but rental gear is subject to invisible damage. Portable control cables get run over by forklifts, and constant patching and bouncing around in trucks takes its toll on dimmers.

One of the live award shows I have done, "The Golden Globe Awards," had such a problem. The show was syndicated, and therefore did not have a big budget. I pride myself on being able to work with limited equipment, but this time it caught up with me. The stage was lit adequately, but I did not have a lot of "toys" to create different looks. I was depending on creating most of the livelier looks in one area where singers were to perform, and otherwise keep the lighting simple. In addition, because backstage space was limited, the fire marshall ordered us to put the dimmers outside the building.

About halfway into the live broadcast, one of the dimmer packs overheated and started flashing on and off. Naturally, it involved a critical lamp: the key light on the master of ceremonies' podium. The first time it went off I thought we had had a blown lamp, so I quickly got a followspot onto the speaker. Only a semi-disaster, but I was saved . . . or was I? When we went back to the MC I put the followspot on the podium. The lamp then came up. What was the problem? A short in the lamp? A bad twofer or cable? Frankly, the last thing I suspected was a dimmer problem, as the equipment was from a top rental company and had worked perfectly up to that point.

The next time we used the lamp, it came on as it should, but then started flashing on and off. Then I knew it was a dimmer problem. However, getting to the problem took several minutes, because of the dimmers' location, and all the while the flashing continued on camera. Now that's a disaster! We shut the dimmer down and finished the show using a followspot to cover the area.

This was not the time to remind the producer that he had cut the budget so tight that there was no money for backup fixtures, which normally would have been built into the design to cover such a failure. I, after all, had ordered the equipment. On live broadcasts, it is considered SOP (standard operating procedure) to double hang areas like the MC to guard against a critical key lamp going out; I could have saved a couple of lamps from other areas and used them to back up the MC. I did not.

Computer Problems   My final example deals with computers. Now, I pride myself on being an early advocate of computer lighting control. I believe I was the first to use one on a concert tour with John Denver in 1974. But failures do happen.

My most recent failure came at a corporate show for a major auto manufacturer. We had rehearsed for a week, two shows were behind us, and we had one performance to go. After a quick run-through in the afternoon, we were ready for the audience.

I was sitting next to the console reading when my boardman returned twenty minutes before the curtain was to go up. As I looked up to greet him, I saw both of the video monitors go blank. My first thought was that we had lost power. We checked it, and the power was okay. I reset the switch, the screens came back on and then we watched the computer go through its internal diagnostic program. The show program came up, and then went blank again. We went through the check and diagnostic program once more, and again the screens went blank, only this time smoke came out of the back of the console.

Now we *knew* there was a major problem. While the operator opened the board, I ran for the telephone to get a backup console on the way. Then I went looking for the producer to tell her we were dead in the water, so to speak. After the color came back to her face, she asked the obvious question: "What can we do now?" I calmly replied, "I'm not sure there is anything we can do."

By this time, the cover was off the console. We found burn marks near the power supply, and a burnt wire lead. We replaced the wire and the computer was back on line—with no show program! The power failure had caused an electronic spike that destroyed the disc drive module. The backup console arrived, but it wasn't a match, so at this point, I had to make a decision. Should I go to the new console and reprogram the show from scratch, or stay with the one I had? I decided to use the original board. By reprogramming the soft patch and reassigning channels to the 24 submasters (and thank God the board had that many submasters), I felt I could at least do the basic show looks. To reenter all 137 cues would have taken too long, so while the board operator entered the dimmer-to-control channel patching information, I went to work laying out the submasters so we could run the show on manual. We brought the curtain up 40 minutes late. That final show turned out to be better than the first two.

However, the problem never should have happened. I should have required a computer with dual power supplies, or a standby duplicate console. It's easy to say that in all the times I had used this particular board nothing had happened. I failed by falling into the *it never happened before* syndrome. There is always a first time.

Being Prepared

Early in my training as a designer, I learned a wonderful lesson from Dr. Sam Selden, author and for many years chairman of the Theatre Department at UCLA, who came to Southern Illinois University as a visiting professor. I was assigned to be his stage manager on a production of *Peter Pan*. Things had been going pretty well, and I was feeling cocky when he came up to me and asked what I intended doing if a particular hydraulic lift did not come up on cue. I hesitated, and he told me to come to him later with three solutions to the problem. It was a good point: we should constantly be considering *what ifs*.

I'm not saying we should or could build redundancy into every part of every system, because that isn't economically feasible. But we should always be prepared for the worst, and learn from our misfortunes. The greatest problem is being unprepared.

II

Equipment Designed to Travel

10

Technical Innovations

The technical innovations that concert lighting is making in theatre, television, and other lighting media are only minimal in the area of radically new products. Generally, the advances involve standard theatrical items that have been modified or repackaged. Like legitimate theatre, concert tour lighting borrowed from any market or service that had something that could be adapted, modified, or used to its benefit.

This chapter gives an overview of the innovative concepts and products used in the concert lighting business. This section is by no means meant to represent all of the innovative ideas being supplied to the touring industry by groups as diverse as small touring companies and big manufacturers.

Subsequent chapters in Part II are devoted to the three major innovations: trusses, lifts and hoists, and moving lights. This chapter will focus on some other areas of equal importance to the technological developments advanced by the concert business.

Consoles

The advances in console design cannot be overlooked, but it would be unfair to say that they came about purely because of rock and roll demands on the established manufacturers. It is true that the complex manual boards of the 1980s, with flash buttons (instant on buttons), chasers (grouping to auto-timed changes), and pin matrixes (shunt boards allowing one fader handle to operate random dimmers), were not available as standard items in the early 1970s. Often the designer would be the one to modify existing consoles. Small concert equipment suppliers did the same. Later, British concert companies started making their own boards and they still lead the field. The early board by Electrosonics (a British company), which was based on a layout concept I brought to them (see Figure 10-1), was probably the first mass-produced board to use flash buttons and pin matrixes. It was aimed specifically at the concert market.

At about this time (around 1972) the major manufacturers were gearing up to bring out computer-based consoles. They seemed convinced that there was no market for a more complex manual console, but then they did not see rock and roll as being a market—yet. Not that the introduction of computers was wrong, they just did not meet the concert designers' needs. Computers are finding use in some rock and roll tours as stand-alone units, such as the Kliegl Performer series and the Light Palette by Strand, especially with MOR acts. But in large manual boards such as the Avolites boards (see Figure 10-2), we find multilevel functions with complex features that allow instant manual control of complex cues. All of their boards feature the now standard flash button array and pin matrix, but some add a computer base to

Figure 10-1 Electrosonics' "Rock Board" console
This is one of the earliest mass-produced lighting consoles designed specifically for tour use by the designer (myself) and manufacturer, Electrosonics Ltd. (Photo by James L. Moody.)

Figure 10-2 Avolites' console QM-500
This is a state-of-the-art concert lighting console that is a synthesis of manual and computer consoles. (Photo by Avolites Production Company Ltd.)

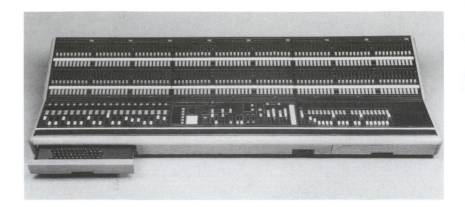

give 400 discrete memories on twenty faders. One unit comes with an alphanumeric display that is available in several languages. The chase mode offers 20 memories in over 1500 steps. A smaller board called a Leprecon is also popular. It offers limited chase but has flash buttons and a matrix, and is popular with clubs and rental houses.

The main reason that most concert lighting designers have not gone over to fully computerized consoles is the hands-on ability to activate flash buttons and faders in time to the music. Designers say they need the manual control to feel a part of the music and they do not seem to get that feeling when only touching one button to execute complex commands. What is sacrificed in accurate repeatability in cue sequencing is made up for by the feeling of making a live contribution to the creativity of the show's mood and tempo.

Dimming

Dimming is an area where packaging, not electronic innovation, was the real advance in the early years of touring. In the 1970s the established manufacturers were just starting to market compact groups of six to twelve dimmers in a portable unit. What the manufacturers did not conceive of was mounting input connections as well as output strips directly to the dimmer packs so that there was much less to assemble each time the dimmers were set up.

Distribution systems with low voltage control patching came next. This enabled the designer to assign one or more dimmers to a single channel on the console. Later, electronic or *soft patch* systems appeared, using a computer to assign these control circuits. In several cases this is done in the console, such as with the fully computerized consoles already mentioned. Some British builders put a "banana peg" cord-type patch bay into their dimmer racks.

The placing of several of these *package dimmers* in a single, castered road case meant that 24 to 72 1 kW dimmers could be connected and ready to operate in a fraction of the time it took to stack and wire each pack individually. In Figure 10-3 the front view (left) shows 72 1 kW dimmers while the rear view (right) has 24 multipin output connectors for fast setup. The pin patch for dimmer-to-control channel assignment panel is in the upper right of the rack. Shipping damage is also considerably reduced with this arrangement. However, to make dimmers "roadable" we often must take them apart to add lock washers and epoxy to keep the components from rattling to pieces in the truck.

U.S. dimmers were being offered in a wide range of capacities: 2

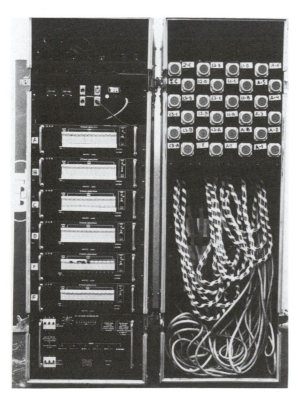

Figure 10-3 Packaged portable dimmer rack (built in England)
(Photo by Avolites Production Company Ltd.)

kW, 2.4 kW, 3.6 kW, 4 kW, and 6 kW. A big advance came with the British manufacturers concentrating on 1-kW dimmers. That allowed each PAR-64 (99 percent of all the fixtures on the road are 1000 watts each) to be patched without twofers or multiple outlets. Now the low voltage patch could take over and assign the dimmer and associated lamp to any one or more control channels. The 1-kW dimmer also meant that more circuits could be housed in a single, movable rack with main circuit breakers, and so on, fully enclosed in one unit.

One such dimmer rack, shown in Figure 10-4 and built by Sundance, uses Skirpan 2-kW and 6-kW dimmer modules and a Rual slider patch system. The rear view (left) shows six Pyle-National multipin outputs. The slide patch assignment panel is above. Designed with built-in breakers and hot bus, the slide patch makes it easier to assign loads and eliminate patch cables. The three 60-amp connectors are the direct output for the 6-kW dimmers. The mini-pin patch in the upper right assigns dimmers to control channels. The front view (right) shows the five modules of 6 dimmer and one module of 3 dimmer modules that are the heart of the system. The right side contains the power section, with digital meters monitoring all phases of both amperage and voltage. A 250-amp, three-phase main breaker and control cable connectors complete the panel.

The latest push is into *high density dimming*. Because the big manufacturers were again behind in the total packaging of dimmers with the ancillary gear, they focused on the research and development of this miniaturization of components. The concert field is currently dominated by small manufacturers of dimmers that provide fully integrated systems in a road case complete with main breaker and multicable connections. The two ideas can come together very nicely and only time will tell if the concert designers and suppliers will return to the mainstream manufacturers for their dimming needs. The current dominance of the market by British manufacturers is mainly because they are directly involved in touring and know firsthand what works for the road. They are only seeking sales to other markets as a secondary income opportunity, whereas the big manufacturers look to installations for their bread and butter. When Sundance was beginning, we made deals with the established dimmer manufacturers to use their main components and let us do the packaging. One example of that is the Skirpan dimmer rack (Figure 10-4).

Figure 10-4 Skirpan/ Sundance dimmer rack
(Photo by Sundance Lighting Corp.)

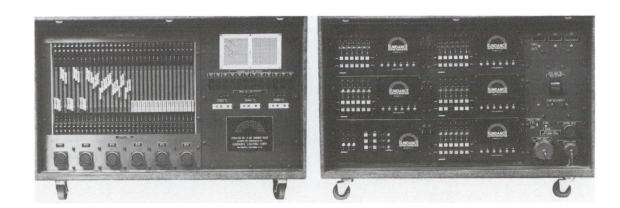

Multicables

The use of portable, flexible multicables may not have been discovered by rock and roll touring, but they have certainly been refined to become a highly efficient and safe addition to the touring system. A multipin connector, such as is made by Cinch-Jones, Veam, and Pyle-National, was the only other item needed to make quick work of cabling.

The standard was the rubber jacketing for cable; other jacketing materials came later. A whole new market for cable manufacturers was opened. The weight of rubber and the fact that it is not very flexible when you have twenty-two to thirty #14 stranded wires inside made concert lighting companies look for alternatives. The British, again, led the field because they had different electrical codes and voltages.

Although not approved by the National Electrical Code for portable use, the fact is that almost all concert multicable has a jacketing of neoprene or PVC. A product called Cranetrol was first used in 1977 by my company after seeing it on a large construction crane. It had the neoprene jacket that could withstand severe heat or cold and yet not get brittle and crack as rubber does under similar circumstances. It is about half the weight of rubber, is more flexible, and can be purchased in colors. The popularity of multicable has found big supporters in installations as well. Multicable used as a "nonpermanent" installation is quicker and more flexible for many theatre situations. Certainly the bus and truck tours of dance groups and theatre can see the advantages of weight reduction and smaller, easily coiled cables. I have seen more and more semipermanent television studios taking advantage of these cables also. If you have ever spent hours taping bundles of #12/3 cables together you will know why the "Hod" will not be missed. To carry the cables, the old wooden box with steel casters that gave you splinters and did not roll very well quickly disappeared from the road. Modern road cases are built of PVC-coated plywood or cabinet grade plywood with indoor/outdoor carpet (see Figure 10-5).

Fixtures

Because power availability was a very serious problem in the initial days of concert lighting, a more efficient light source was needed. The Fresnel and the older plano-convex spot just did not have the lumen output needed to project heavily saturated color onto the stage. Also, they did not go on the road well. They were heavy, rattled apart, and did not pack easily.

The PAR-64

The answer to the problem was found in a relatively new lamp source being used on location shooting for film and television. The PAR-64 as we know it was a modification of the *Cine-Queen* fixture introduced in the early 1960s by Berkey-Colortran (see Figure 10-6). Several other manufacturers had made copies but it was the modification of a snoot to hold color far enough out in front of the lamp so it would not burn (made by Altman Stage Lighting at the urging of Bill McManus) that allowed it to become the standard for concert lighting. The lamp had been housed in a can with nothing more than a yoke and power leads.

Figure 10-5 PVC-type cases
(Photo by Excalibur Industries.)

Figure 10-6 Cine-Queen
The Cine-Queen, shown here with intensifier ring, was the original housing in wide use for the PAR-64 lamp. It was used for film and television production where color media was not of concern. (Photo by Colortran Inc.)

Figure 10-7 Ray Light
(Photo by Creative Stage Lighting Co., Inc.)

The PAR-64 had been used as a punch light by film in banks of six or nine to replace the Brute arc-type lamp that required 220-volt power and an operator to run it. (A punch light is a fixture with a highly concentrated beam that allows long throws.)

The PAR (Parabolic Aluminum Reflector) lamp had the highest lumen-per-watt rating of any lamp then developed in the 3200 Kelvin range. This translated into a fixture that had no moving parts, no lens to break, and lots of light! Heaven had arrived for the concert designer. In a media where bravado was a key element in design, the PAR-64 meant low initial purchase price and long road life. The economics were clear: a PAR-64 housing cost about half that of a Fresnel and a third to a fourth that of the ellipsoidal reflector spotlight.

This sealed-beam lamp offers a range of beam spreads from very narrow to narrow to medium to wide flood. The ACL (Aircraft Landing Light) can also be fitted into the standard PAR-64 housing. It produces a very, very intense narrow beam of light. The drawback is that it is a 24-volt lamp. While four can be ganged in series so that our 120-volt dimmers can handle them, if one goes out you lose all of the group. You also give up some individual control. And yes, they really are for aircraft. I remember an aircraft supply company calling us and asking how big a fleet of aircraft we had because we ordered so many lamps we had completely depleted their stock.

A recent advance is called the Ray Light (see Figure 10-7). The reflector fits in the standard PAR-64 housing but uses the DYS lamp, which is a small, minipin base lamp at 600 watts. It throws a very concentrated beam, very similar to the ACL, but operates on 120 volts. There is a 30-volt version of the lamp at 250 watts. Reflectors are also made for the PAR-56 and PAR-36 as well.

I find it interesting that theatre has chosen to ignore the PAR-64, often saying that it is an open-faced fixture and therefore an uncontrollable source. Nevertheless, the scoop and beam projector are lensless fixtures commonly used in theatre. Along the same lines, television would have a hard time doing without the scoop, broadlight, and Cyclight, which are also lensless fixtures. See Chapter 21 for an unusual application for open-faced fixtures.

Fresnel and Leko

I will leave it up to the standard theatre lighting textbooks to fully discuss the lighting characteristics and quality of light you will get from the standard fixtures. However, I would like to correct the misuse of a word I have often seen on light plots and in contract riders. The word is *leko*, often misspelled "liko" or "leco." This is a contraction of the names of the two men credited with the introduction of the ellipsoidal reflector spotlight, Ed Levy and Edward Kook. It is the registered trademark of the old Century Lighting Company, now Strand, for their version of the fixture (circa 1932).

While the Fresnel and leko have not been totally abandoned, I estimate their use at less than 5 percent of all the fixtures found in concert lighting. Other casualties of the concert market would be the scoop and the borderlight.

Cyclight and Farcyc

Concerts have adopted one television fixture: the Cyclight is widely used for lighting backdrops. The design of their reflector and the light output, if not their size, have gained them favor. They take up a lot of room and do take time to set up, but the advantages outweigh their shortcomings. Several versions of these are made. For floor mounting, a series of four to twelve lamps in three to four circuits can be used. Overhead mounting favors the Farcyc, which is a square of four lamps, and was first used at the BBC studios in London.

Followspots

Truss-mounted followspots are a rather common device for providing concentrated light on a moving subject. Coupled with a mounting bracket and chair with a safety belt, the followspot can be placed anywhere the designer chooses without worrying if the building will have positions where needed.

The smaller HMI units are popular. Newer versions have introduced the HTI lamps (metal halide short-arc lamp) in such models as Ultra-Arc's Mighty Arc II™ and Lycian's SuperArc 400™ (see Figure 10-8), which both use a 400-watt HTI source. These lamps are only about thirty inches long and put out better than 300 fc at the normal throw from the truss—a very hot source that is very effective in concert lighting design. The HMI and HTI sources are 5600 degrees Kelvin, so adjust your color media to account for the bluer light.

There are also several remote-controlled followspot units on the market. These were designed to remove people from the trusses, and certainly it is a safety advantage. Morpheus Lights has the Cue-Spot™ and the French company Cameleon produces the Telescan™ as the remote units.

Smoke and Fog

A very important ingredient in concerts, especially with rock and roll and heavy metal bands, is some type of medium that allows the beams of colored light to be seen by the audience. This is especially important when Air Light or moving lights are being used. The patterns created over and around the players would not work if it wasn't for some means to accent the beams (see Plate 1).

Enter *theatrical smoke*, or, more correctly, what *appears* to be smoke, not a smoke created from combustible materials. At first the only semisafe way to produce a controlled release of a smoky substance

Figure 10-8 SuperArc 400 followspot
This followspot is very compact at 35½ in long and very powerful. Using a small HTI source it produces 1500 fc of light at 40 feet (spot position) and is extensively used as a truss-mounted unit in touring. (Photo by Lycian Stage Lighting.)

was borrowed from film. The Mole-Richardson fogger used an oil-based liquid that was heated until it vaporized into a cloud of smoke for special effects of night fog, simulated battle scenes, or fire effects.

This proved to be very irritating to singers and the search was on to find a more acceptable product. It didn't take long for people to experiment and develop products such as those introduced by Rosco Labs. Rosco was even given an Academy Award in the Scientific and Engineering category in 1984, "for the development of an improved nontoxic fluid for creating fog and smoke for motion picture production."

Since then several other products have hit the market. The Great American Market offers some with scents to take away some of the oily odor. These are much better but they all have one problem—control. The units that heat these liquids have also come a long way. Some have timers to allow for a controlled release and most have remote controllers, but the problem comes after the release.

Smoke is lighter than air. Therefore, air currents take the smoke wherever the wind or hotter, rising air wishes to go—not necessarily where you would like to keep it: on stage. Placement of foggers can be a real science. Remember that buildings often do not have the air conditioning or heating on until the audience is about to enter. Besides doors opening and closing, even the heat from the lighting will influence where the smoke is carried. Experimentation, as well as bringing along your own fans, works most of the time.

There is one other way to produce theatrical smoke. The use of some pyrotechnical devices will create smoke. Use caution, as most states require a licensed pyrotechnician to do this. There is one product that is used most often in television and films. It is called Spectrasmoke™ and is made by Tri-ess Science Inc. in Glendale, California. Another version is produced by Luna-Tech in Huntsville, Alabama.

In 1983 the Hazardous Substance Act went into effect. When Tri-ess was given approval in 1984 to sell the product, the U.S. Bureau of Explosives stated in its report, ". . . recommended that the materials are classed as Flammable Solid, N.O.S. under DOT regulations." Tri-ess feels that they are safe for theatrical use and indeed they do a brisk business in Las Vegas and the Hollywood community. The solid comes in powder, shaped into what looks like a big sugar cookie form, and in grenades. Several colors of smoke are also possible. Methods of ignition depend on the form you purchase. The company states that during ignition and combustion, it burns at a low 170 degrees and will not ignite a piece of paper placed under it. The advantage over the liquid is that as this burns, it releases particles that tend to hang in the air better. It is not designed for use as ground fog.

Whatever you use, always be aware of smoke detectors and alarms, especially in hotel ballrooms. They can be set off by any of the smoke-producing agents. Second, even though all are said to be nontoxic, all carry warnings that irritation to eyes and lungs can occur in sensitive persons. Be sure to check with the artist before using these methods to generate smoke. In many states and municipalities these are considered a pyrotechnic device and you will be required to obtain a permit.

The Need for Better Equipment and the Solution

It would be easy for the lighting manufacturers to say that these advances would have happened eventually, but they cannot deny what rock and roll provided: ready cash, no-bid purchasing. Theatres often take years in planning sessions developing specifications. Most touring lighting companies have one person, the owner, who makes all purchases, which often run as high as a quarter of a million dollars. Rock and roll brought to the manufacturers a demand for prompt, efficient products and service. It has been a boon to all media because rock and roll stimulated the equipment manufacturers where it counted—in their cash flow.

In all the areas we have looked at in this chapter one single need was present, the need for "roadability." The equipment must be able to withstand the constant bouncing around in trucks all night, and be easily set up, used, struck, and packed back in the truck.

Now that concert lighting has a proven solid base, it is being discovered by a wider market. The use of any and all of these products can and often does have application in other media. Our continued livelihoods as designers and technicians require that we stay constantly on the alert for new ideas and techniques to better solve our design and production problems. Chapters 11, 12, and 13 will give you a close-up view of three of the areas I feel will have the most influence on the wide range of "theatrical" lighting fields over the coming decade—lighting trusses, lifts and hoists, and moving lights.

11

Lighting Trusses

The use of "found space" for a theatrical production is not new. Barns, outdoor arenas, and all manner of multipurpose rooms have been used for stages. The elaborately equipped buildings designed for large symphonies and opera as well as those built specifically for drama in the early twentieth century are still in the minority. Because nearly all of the large theatres were built with stagehouses of a similar size and design (except for thrusts, et cetera), it was relatively easy to bring in a touring opera or theatrical production to those theatres; most of the buildings had permanent lighting pipes in neat rows to provide lamp support. But what about unequipped buildings?

The First Trusses

Rock and roll concerts were deemed "not artistic enough" for many city and college theatres, and were generally relegated to the school's gymnasium. Where do you hang lights in a gym? Bring the structure to mount the lights with you. Create a performance area in the found space. Constructing *portable* units that could be trucked easily from show to show was the goal.

Portable units also needed *time-saving* features. To make money, the recording artists traveled and performed in a different city almost every night. A play usually has a run of a week or two with possibly a day or two of rehearsal to adjust to the particular theatre. For concerts, speed was important. The solution to this new problem was hit and miss at first, but as more people took it seriously new design ideas emerged. Creative people took on the challenge and a whole new design world was born: a structure, somewhat like a bridge member, that allowed the designer to place lamps overhead instead of on standing poles or trees.

The structures themselves have no historical precedent in theatrical design. The first truss for touring music was designed by Chip Mounk and Peter Feller with Bernie Wise for the 1972 Rolling Stones tour. The lamps were not left in place when packed for traveling. It was ground supported. In 1973, a box truss with fixtures mounted for travel inside the truss was designed by Bill McManus with Peter Feller and Bernie Wise. It was the first (hung) truss grid and measured 50 ft × 28 ft flown using CM hoists™. It was designed for the Jethro Tull tour of that year, known as "The Passion Play." Other young rock and roll companies such as Showco, Tom Fields Associates, and See Factor were right behind them in developing very individualistic designs that with each new tour brought a new idea to the scene.

Triangle Truss

Triangle trusses fall into two groups: commercial towers and specialty trusses. Those built commercially as antenna towers are available in several widths and tubing sizes. Because these are designed primarily for vertical stress, they usually cannot be used horizontally without some bowing or sag. When flown, however, the pickup points can be placed strategically so as to eliminate this problem (see section titled Spans below).

There are commercially made versions available now that copy this type of construction but are designed to withstand the horizontal stress and loads we place on them for concert lighting (see Figure 11-1). Some of these specialty trusses are constructed of a heavy chrome-moly, but most are of a lightweight aluminum construction. Their initial cost is low at under fifty dollars per ten feet. Repeated tightening of C-clamps will compress the tubing, making their useful life short. Because more development has been done on these specialized trusses, such as the one shown, the safety factor has been increased. However, use extreme caution in using commercially made products that are not specifically designed for horizontal stress.

Two types of design are used in triangle trusses. The first is a solid triangle in 10 ft lengths. Widths range from twelve inches to thirty inches. The second design is constructed with a hinged joint at the top and a spreader bar attached on the horizontal side (see Figure 11-2). Removing the spreader bar allows the sides to close for compact storage.

In both cases, fixtures must be attached once the truss is supported in position. For a onetime production, this may not be a problem; but the following disadvantages should be considered before using these trusses.

1. The fixtures can be focused safely from the ground only by ladder.
2. Fixtures must be attached and plugged each time the truss is set up, a time and labor disadvantage.
3. Because fixtures must be attached to the triangle, the usual method is individually via C-clamps or hanger straps on either end of a six-lamp bar. Adding sixty to two hundred C-clamps, at about two pounds each, is an undesirable weight gain.

Figure 11-1 Triangle truss, rigid
A basic truss of steel tube construction with ¹⁄₁₆ in wall thickness, designed for a maximum length of 48 feet and a working load capacity of 1,200 pounds. (Photo by Concert Lighting Systems Australia Pty Ltd.)

Figure 11-2   Triangle truss, folding
(Photo by SeeFactor Inc.)

A few modified triangle trusses do travel with fixtures inside the structure, but they are not the rule. Figure 11-3 shows a unique approach that allows a row of lamps to be transported fully wired and colored, with extra fixtures able to be added on-site.

Square or Box Truss

Square or box trusses come in a variety of configurations. Some, even though constructed as rigid units, still use hanger straps to mount the fixtures to the structures when on-site. Two such box trusses are shown in Figure 11-4, which can be used not only to mount lighting fixtures, but to rig drapery and sound systems as well. The center unit is a six-way corner block used to join trusses in several configurations.

Other box trusses are large enough to semipermanently mount the lamps internally to Uni-Strut bars or T-connectors. The prerigged truss

Figure 11-3   Triangle truss, modified with lamps
(Photo by T.E.R.I. Inc.)

Figure 11-4   Box trusses

Front, 12 in square unit; *center,* corner block; *rear,* 20½ in square unit. (Photo by James Thomas Engineering Ltd.)

section shown in Figure 11-5 is 91 inches long and 20½ inches square and shows two lamp bars of six PAR-64s in the transport position. Pins are released and the bars extend below the frame when in use.

Trusses can be built to accommodate single or double rows of fixtures. Figure 11-6 shows lamp bars in the lowered position. Note the casters for easy movement around stage before rigging. The side view in Figure 11-7 shows the offset mounting used for a better focus angle. The lower left bar removes after transport for unobstructed focus. Here the casters are on the back of the truss so they do not hang below.

Figure 11-5   Double-hung box truss (transport position)

(Photo by James Thomas Engineering Ltd.)

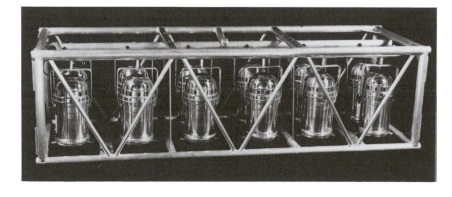

Figure 11-6   Box truss, extended lamp position

(Photo by Pete's Lights Inc.)

Truss Design

There are as many variations on the theme of truss design as there are companies designing and building them. Each company feels its design offers the solution to a particular problem, such as speed in setup, strength, lamp capacity, or the method in which the sections are interconnected. For flexibility in design application, remember that whether you mount lamps internally or not, the four sides of the box and even the ends are available for mounting, so bottom corners can be used as well as top corners.

A major advantage of truss use is the flexibility afforded by corner blocks and angle blocks to connect truss sections into shapes. The concert designer's truss layout is discussed further towards the end of this chapter.

Because of the basic structure of a triangle, there is not much room for creativity in internal design. The box truss, however, has spawned some highly creative modifications to its basic form. One design has a cable tray in the top so that cables need not be disconnected for packing in separate boxes (see Figure 11-8). Another folds open to create two levels of fixtures, and in the closed position provides its own protective case. In several designs, the lamps extend out from under the box truss when in position so that the structure does not interfere with the beam of light.

When doing location video lighting, the only problem in using trusses is that most are built to accommodate PAR-64 fixtures only, and a 2 kW Fresnel might not fit inside. There are companies that can provide trusses to hold some larger fixtures, but I know of none that hold a standard size 5 kW Fresnel internally. When the 2 kW or 5 kW Fresnel must be used, I suggest the triangle or small box truss with the lamps mounted externally as the easiest solution to a onetime location problem.

A new twist in truss construction is not in material, size, or design, but in tube covering. Until recently most trusses were left unpainted. The silver color of the metal allowed the trusses to be seen when they

Figure 11-7   Double-hung box truss, internal support
(Photo by Sundance Lighting Corp.)

Figure 11-8   Truss stack with cable tray, an efficient design with fewer casters
(Photo by McManus Enterprises Inc.)

were a part of the stage design, but there was not much that could be done to hide them when desired. Now the firm Rainbow Productions has introduced a polyester matte finish for their trusses.

Engineering and Construction

Most of the trusses are built of aluminum tube #6063-T5 in 1 in and 2 in O.D. or H E 30 alloy, and with a fairly heavy inside wall thickness. Chrome-moly is being used less and less although it is cheaper and easier to weld. The added weight, about twice as heavy as aluminum, is not desirable for touring. If the trusses are for semipermanent installation, chrome-moly should be considered for its cost alone. Its other advantage is that it can be welded and aluminum must be heliarced. The welder does not have to be as highly skilled to arc weld as he does to use the heliarc machine.

Although steel welding is less expensive in materials and equipment, and labor is more readily available, steel is seldom used in the United States, largely due to its weight and to the controversy surrounding the employment of amateur or semiqualified welders. This can create a tremendous liability problem that I feel should be avoided. We are working in a highly experimental area here and the potential for a wrongful death or injury suit is substantial.

You can purchase trusses from one of the several companies who are in business strictly to sell trusses and not to design touring lighting. Make sure you are provided with engineering stress information. Because they will not be loading lamps on the trusses, the company cannot be held responsible if you misuse the truss.

The engineering of trusses is critical and is probably the biggest reason why onetime shows should not try to construct their own trusses, but rather lease them from an established concert company. If you are considering construction, I recommend you use only certified welders. Although actual construction time could be about five days for a 40 ft truss, it is essential to have a certified structural engineer design or check your idea. This added cost and time is another reason to lease if your project is short term, or consider purchasing if long term. Be sure to ask the company for a certification of structural stress and load capacity. This can be done by specialized engineering firms and should cost under one thousand dollars, depending on what procedures they use and how far you carry the tests. Figure 11-9 shows the basic procedure used to test the trusses for load.

Spans

As there is a large difference between the strongest and weakest truss, a certificate of load capacity is very important. Moreover, the clear spanning capability of the truss must be determined. Some trusses can only be supported up to a clear span of forty feet. Not only length but the size of the tubing and design of the truss must be considered. Generally, truss sections come in 8 ft and 10 ft lengths. The rental companies will help you fit your design into the lengths they have in their inventory. Even better, contact the company and get a breakdown on what is available before you begin a design.

Now that most trusses are flown, the full forty feet of a standard portable stage is usable. (Ground support lifts reduce the usable width to thirty-two feet.) That is why forty feet is considered the average

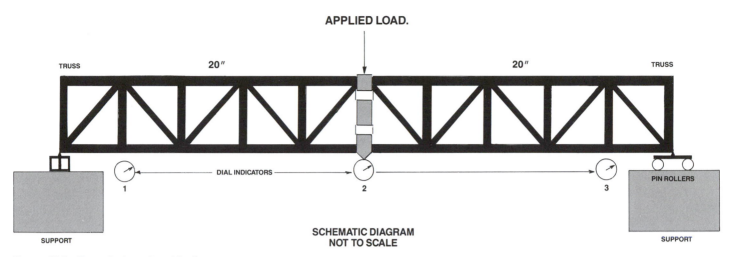

Figure 11-9 Sample truss load test

truss length. Larger productions, however, are now calling for 50 ft and 60 ft lengths.

It is rare to see lengths over forty feet being floor supported. If your stage requires a truss over forty feet wide, it is time to consider a flown system. Why can a longer truss be flown when it can not be ground supported? The solution is found in placement of the pickup points. Bridles (flexible nylon webbing or ¼ in aircraft cable) are composed of two overhead load points of lesser load-bearing capacity being joined to lift a heavier load. They help to distribute the weight evenly. It is common for a 40 ft truss to have two motor or winch pickup points approximately five to ten feet in from each end. These, in turn, will usually be bridled about four to six feet apart. The proper bridle configurations for a given load must be determined by a qualified rigger. The example in Figure 11-10 shows some of the calculations needed to determine the stresses.

I feel flown trusses are safer than ground-supported trusses. Out of necessity, the ground-supported trusses are at the mercy of many factors, such as a stage with uncertain construction that could collapse under the weight. In several reported cases, lighting companies have refused to set up a ground-supported truss because they felt the stage was unsafe.

Electrical Connections

The electrical raceways and cables attached to the truss do affect the structures. There are several methods of getting power to the fixtures; the simplest method is for fixtures to be wired on-site. Another way is to use the old theatrical road-show method of making up hods of #12/3 cables with the ends spread out to the fixture positions and then connected on-site.

The standard electrical raceway takes this method a step further. It is either placed on the truss once on site or attached semipermanently to the truss. The inner connection of the fixtures can then be easily accomplished if the lamps are to be mounted each time, or patched and left for the run of the production if the raceway is an integrated part of the truss. The fixtures can be permanently wired to a raceway

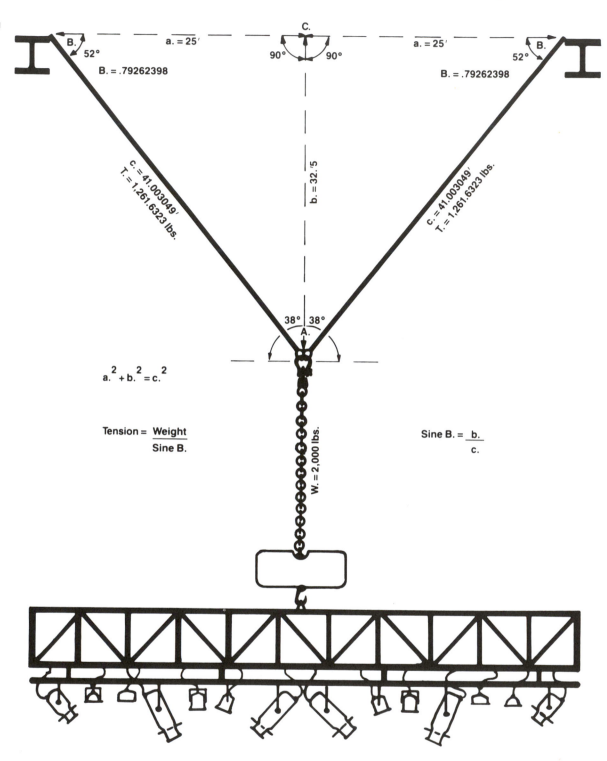

Figure 11-10 Rigging bridle example
(Provided by Rigstar Rigging Inc.)

B. = 52°
B. = .79262398
a. = 25'
C.
90° 90°
a. = 25'
B. = 52°
B. = .79262398

c. = 41.003049'
T. = 1,261.6323 lbs.

b. = 32.'5

c. = 41.003049'
T. = 1,261.6323 lbs.

38° 38°
A.

$a.^2 + b.^2 = c.^2$

$$\text{Tension} = \frac{\text{Weight}}{\text{Sine B.}}$$

$$\text{Sine B.} = \frac{b.}{c.}$$

W. = 2,000 lbs.

as well. This last method, however, inhibits design changes and fixture replacement and is generally considered inefficient for any use other than a straight all-PAR-64 production design.

Lighting Grid Design

A total lighting grid can be formed with one type of truss or with a combination of sizes and designs. It is not uncommon to see a single square truss in the front and on the sides and a double-row truss in the back for the more important back light.

Trusses are joined together in several ways. End blocks with bolts, aircraft fasteners with a ball-lock capture pin, cheeseboros such as are used on scaffolding, and other highly specialized connectors are designed especially for this use.

Some trusses have been designed in such a way that a single-row truss will attach to another to create three or four rows of light. The way in which the designer lays out the configuration of trusses to accomplish the lighting needs is limited only by the load limitations of the lifts, winches, and motors used to place these structures in the air (see Figures 11–11 through 11–16).

The simple tour design in Figure 11-11 consisted of two 32 ft trusses supported by Super Lifts. Note that the backdrop is hung from the truss also. Figure 11-12 shows a unique design that allows cable to be easily routed from truss to truss and gives access for focusing. The lighting trusses in Figure 11-13 were surrounded by a truss grid of drapery that was raised via motors. It created a three-sided house curtain to conceal the stage prior to the performance.

Figure 11-14 shows a design of lighting trusses with motorized mylar mirrors that was used by the rock group Genesis. Spots positioned upstage were aimed into the mirrors and the beam bounced back down onto the performer. The angle of the mirrors was changed to allow the light to follow the performers around the stage. The concert designer for Kiss used a hexagon shape for this show (see Figure 11-15). The truss design was a very important part of the total stage design. Figure 11-16 displays a lighting grid with followspots. In the upper right corner is an Ultra-Arc followspot that provided back light cov-

Figure 11-11 Stage-supported box trusses
(Photo by Sundance Lighting Corp.)

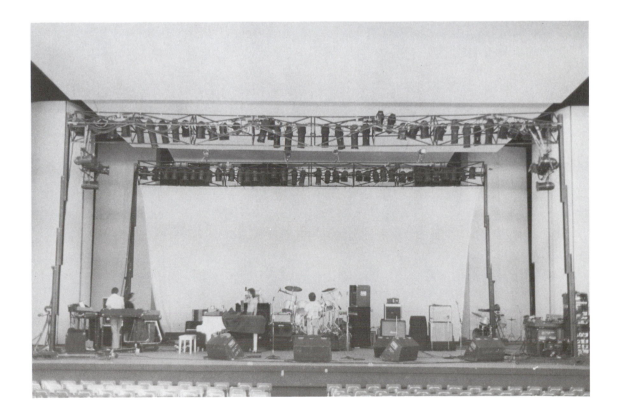

Figure 11-12 Truss grid with walkway
(Photo by Showco Inc.)

Figure 11-12 Truss grid with walkway (Photo by Showco Inc.)

Figure 11-13 Overhead view of flown truss grid (Photo by Sundance Lighting Corp.)

Figure 11-14 Mirror
truss, Genesis show
(Photo by Showco
Inc.)

erage of the artist. Other mounting designs use a seat attached directly to the truss or mount the chair and followspot below the truss for a totally clear shot of the stage.

A designer should also consider the capability of the roof structures of the halls before using a flown truss layout. While the designer is not expected to be a qualified rigger, some understanding of the type of venues and what shows have played in the room previously is a good guide. Consulting a tour rigger before you are committed could save you a lot of problems.

Advantages of Trusses

A truss that has been loaded with fixtures and cables and has been checked in the shop prior to going to the location will give savings in on-site setup time. The structure's efficiency means that less labor is used in the field. This does not mean the main reason to use a truss is to put people out of work. Trusses still require labor to prep them in the shop. They do, however, reduce the producer's on-site costs and often also mean fewer people on the road, thus saving hotel, travel, and per diem expenses. Using trusses can even mean savings of a day or more of site setup time, which translates into possible rental savings; and more important, it may be the only way to get the production to fit into the very tight schedule of the facility.

Trusses provide a very convenient, adaptable lighting system, but the safety element must be stressed again. The problem of getting lamps up to a fixed pipe twenty feet in the air creates a very real hazard. I have ducked out of the way of many falling items during such load-ins. Just the problem of working at this height rather than at ground level should be obvious. Trusses have been in use since the

Figure 11-15   Kiss truss system
(Photo by McManus Enterprises.)

**Figure 11-16   Lighting
grid with followspots**
(Photo by Sundance
Lighting Corp.)

1970s and have proven themselves reliable and efficient in supporting lighting of all types as well as drapery, projection screens, and scenery. Their adaptability for use in films, television, and theatre rigging should be a great boon to flexible and safe mounting of lighting under less than ideal conditions.

One other thing gets attached to trusses—people. A very popular design element is the followspot placed on the flown truss to give creative angles not possible from the house positions. I have two notes of caution. First, never place someone on a ground-supported truss if movement has not been secured via a safety of some type. Second, do so on a flown truss only after the supplier and rigger have okayed the trussing and its rigging for such use.

12

Lifts and Hoists

When trusses were first used as a solution to the problem of lamp support they were, for the most part, ground supported. When they began to be tied to the building's structure, a process called *rigging*, several elements had to be considered.

First, many of the buildings did not have adequate structural supports to hold the added weight. Second, the cost of hiring riggers was prohibitive. As the lighting systems and production complexity grew in later years, there was no choice but to use rigging. This meant that the tour had to restrict itself, in some cases, to larger facilities such as basketball and hockey arenas.

But the market for artists playing in smaller settings was still thriving. How could their shows be expanded, but avoid rigging? The logical solution was improved ground support. There were devices used in the construction trades that could be adapted for touring use. Units were already on the market that allowed workers to change light bulbs and mount materials overhead in buildings.

What follows is a summary of some of the types of lifts adapted for concert use. Not all manufacturers are represented, but every generic type of lift I found is presented, paying special attention to the ones that have received the most use through the years. These items represent savings of thousands of hours in labor expended. They made the difference in doing shows with the lighting available, rather than limiting bookings to theatres with existing structures. These lifts are a key to bringing theatre, dance, opera, and other entertainment to portions of the population that do not have equipped theatres in their communities.

Nonhydraulic Lifts

Genie Tower

The genie tower is a commonly used lifting device that was originally designed to hold commercial lighting fixtures in position during installation. It was the first widely used lift for concert lighting. Fully extended, the unit has a maximum height of 24 feet. It uses compressed air to operate a series of aluminum columns. Maximum lifting capacity is 300 pounds up to 20 feet, 250 pounds up to 24 feet.

The advantages are quick setup and compactness in shipping. One disadvantage is that the lift can stop only in a fully extended position unless restraining cables or chains are used. These cables also prohibit the columns from rotating freely. Another disadvantage is that improper leveling causes air to leak from the seals inserted between column sections; when this happens the tower loses pressure and the columns compress. Use of this unit for support of a truss is not recommended.

The most widely used application for the unit is to lift twelve to eighteen fixtures, supported in a frame or on pipes placed in the holes in the head of the unit (see Figure 12-1). Many crews also box the

Figure 12-1 Genie Tower with lamps
(Photo by Sundance Lighting Corp.)

tower and lamps in a protective hard-sided case or open steel frame-work (shown) for transportation and quick setup.

Vermette Lift

Originally, the Vermette lift was used on construction projects to put air conditioners in place on the second floor. This tower is constructed from lengths of 4 in square steel tubing. The sections of square tubes can be taken apart for travel. A head block and hand crank lift a forked arm via a ¼ in aircraft cable. Models come in 12 ft and 20 ft units. There are leveling devices in the legs and the lift can be used to lift fixtures on a frame or as a support for a truss. The maximum load capacity is 500 pounds. The 20 ft model requires guy-wires to keep the column straight (see Figure 12-2). A considerable bow can happen if the column is improperly balanced. A major disadvantage is that if the cable breaks, there is no safety method provided to keep the forks from falling.

Genie Super Tower

The Genie Super Tower operates somewhat like the Vermette lift. Although it uses wire rope over a series of pulleys, it has a unique advantage. It has a safety braking system approved by the Occupational Safety and Health Administration (OSHA). Another advantage is that the columns nest inside one another and are pulled out as you crank the forks up (see Figure 12-3). The telescoping sections allow it to be used at less than the maximum extension.

Two models are available: 18 ft and 24 ft versions, each with a load capacity of 300 pounds. These units have two types of base configurations, both with excellent leveling jacks. Although the Genie Super Tower has advantages over the Vermette lift, the cost is almost double. It was developed by company president Bud Bushnell and Bill McManus of McManus Enterprises. It must be noted that Genie Industries was one of the first outside companies to take a real interest in concert touring needs.

Air Deck

The Air Deck is a compressed air-operated lift that is an adaptation of the Genie Tower. Essentially, it takes three of the Genie air columns tied together with a basket on top to create a lift for a person (see Figure 12-4). Normally not used in lifting trusses, the Air Deck is very popular as a followspot platform and as a focusing platform. It has operating heights of 24 and 36 feet. The unit weighs in at 351 pounds and has a traveling height of 7 ft 5 in. It is marketed by Upright Scaffolding Inc.

Hydraulic Lifts

Several oil-operated ram lifts are available. They are also borrowed from the construction industry. Although models can go very high (45 feet), the weight and size usually prohibit using them for touring. Load capacities of 500 to 1600 pounds are available. They have the same disadvantages as a Genie Tower: if the load is not directly over the column, the seals are broken and the unit can slip. Although very difficult to travel with, I have used them on three occasions. It is a matter of finding the item that suits your needs, not of being restricted to what you see in theatrical books.

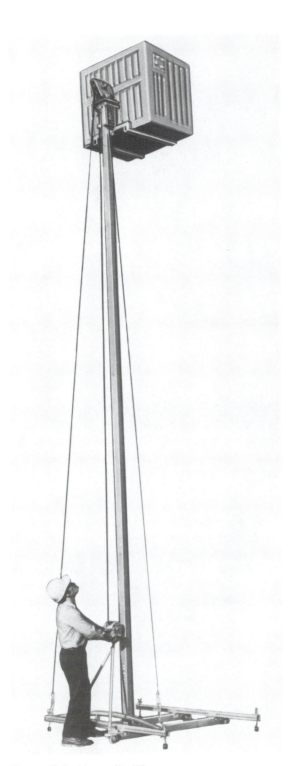

Figure 12-2 Vermette lift
(Photo by Vermette Inc.)

Figure 12-3 Genie Super Tower
(Photo by Genie Industries.)

Figure 12-4 Air Deck
(Photo by Sundance Lighting Corp.)

Flying-Tiger

The Flying-Tiger is a self-propelled, electro-hydraulic elevating work platform that has a payload of 1000 pounds and goes up to 27 feet in the air. It is of the scissorlift variety and weighs in at 1200 pounds dry weight. The larger version is called a Flying-Carpet™ and extends to 36 feet with a larger working platform.

Tel-Hi-Scope

The Tel-Hi-Scope is only one of a large group of hydraulic, single-ram lifts that can be AC operated on external power or operated with batteries mounted on the unit. Models come in 24 ft and 45 ft versions and can carry loads up to 500 pounds. The 24 ft unit weighs 1500 pounds and the 45 ft model weighs 3900 pounds. The obvious first problem is weight. Other disadvantages are: they cannot be disassembled; they have a minimum clearance height of 6 ft 7 in; and because of the oil reserve, they cannot be tipped on their sides. Double-ram units can go higher and carry up to 4,000 to 5,600 pounds, but the units can weigh up to 8,000 pounds.

Special Units for Lighting

Because of the unique problems of touring, several of the concert lighting companies have developed their own lifts. Although some companies are in the market to sell their lifts, most do not. The units discussed below only represent a few of the ideas.

Show Tower

The Show Tower is a self-erecting ground support unit capable of lifting one ton to a height of 28 feet. It uses a CM Lodestar™ chain hoist for power. The base is constructed of carbon steel and the mast is of 6061 aluminum alloy. This design is not patented and it has been widely copied both in the United States and Japan.

Thomas Tower

While Thomas Towers also use CM hoists to raise the trusses, the hoists ride up with the truss rather than being secured at the base as with the Show Tower. The lifting capacity of one ton is the same, but the Thomas lift is two feet taller than the Show Tower at 30 feet fully extended. The towers are erected after the truss is assembled (see Figure 12-5) and are placed inside the box of trusses (Figure 12-6). Note that six towers in Figure 12-6 are lifting the total grid. The motors are not synchronous, but travel at a close enough speed to keep the grid reasonably level. Control units allow for a single motor or any combination of motors to work together.

The Thomas tower is twelve inches square and the total system is designed to lift a box truss with a maximum span of forty feet. There is an optional base with outriggers that allows a tower to stand alone (see Figure 12-7). The photo shows the base detail with a Lodestar motor attached. The light frame holds thirty-six PAR-64 lamps.

Crank Versions

There are a few towers that use hand cranks. Morpheus Lighting uses a flat ladder with a base and cables much like the Vermette except there are no lift forks (see Figure 12-8). The Morpheus lift is easily

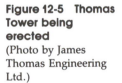

Figure 12-5 Thomas Tower being erected
(Photo by James Thomas Engineering Ltd.)

Figure 12-6 Thomas
Towers with full truss
grid
(Photo by James
Thomas Engineering
Ltd.)

Figure 12-6 Thomas Towers with full truss grid
(Photo by James Thomas Engineering Ltd.)

climbed for focusing work. Stabil-lift™ is another hand-crank lift; it uses a triangle tower, but with much the same result.

Rigging and Hoists

There are a very few ways to accomplish rigging the trusses. The devices made for this purpose are very difficult to find. Few companies who make chain hoists have entered the concert market. The other methods of rigging are ones developed by individual companies and have not gained wide acceptance.

Chain Hoists

On large systems, the circus approach to rigging is taken. By securing cables to the ceiling support beams of the building, a truss or grid can be lifted and suspended. The most widely used method uses a Lodestar motorized chain climber (see Figure 12-9). For concert purposes, the motor is used in an inverted position; special modifications must be made for this application, however, and they should only be done by a factory-authorized agent. CM Hoists, the manufacturer, is very aware of the touring applications and works closely with road technicians and rental houses to ensure proper maintenance. A certification program is also in effect through the manufacturer.

The load rating of the hoists in this series is from ¼ ton up to 5 tons. The biggest drawback is that the chain can fall out of the collector bag, and there have been reports of slippage or brake release when the motor is disengaged. Many of these problems are due to poor operator handling and should not be interpreted in any way as the fault of an unsafe product. CM Hoists is a very concerned and helpful company that has always worked closely with the touring market. The

Figure 12-7 Thomas Tower, free-standing with light frame
(Photo by James Thomas Engineering Ltd.)

Figure 12-8 Morpheus lift
(Photo by Morpheus Lights Inc.)

person using the device has a great responsibility to make sure the device is checked frequently and used properly at all times.

How the hoists are attached to the trusses is up to the user. The placement and size of each hoist is determined by the rigger. Figure 12-10 is a photo of a flown lighting system with at least eight chain motors visible. Note the bridles on the two upstage points. In Figure 12-11 the motor to the left is bridled to the lighting truss to distribute the lifting load to two points instead of one. Note the "horse bucket" that catches the chain as the motor climbs up the suspended chain.

Stanal Method

Although several companies have attempted to build specially constructed motor or lift devices, the most successful method has been the one developed by Stanal Sound Inc. Their method does not use chain hoists; it places a motor on a steel grid and uses a drum to wind up the aircraft cable that has been attached to the ceiling supports of the building (see Figure 12-12). The motor stays attached ot the platform it is lifting while the two cables go through pulleys to a center double-sided drum attached to the motor, which winds the cables in opposing directions; thus the system is self-leveling. Depending on the application, the aircraft wire rope used is ¼ inch to ¾ inch.

The advantages of this method are not in lifting power, but in convenience and safety. It is faster and lighter to pull up 50 feet of ¾ in aircraft wire rope than 50 feet of ⅝ in chain. The aircraft cable is relatively inexpensive and can be replaced frequently when any kinks or burrs are found. All the ¾ in aircraft wire rope, bridles, and clamps needed to rig ten to twelve points can be carried by one person as excess baggage on an airplane. To attempt the same thing with a chain hoist would be impossible. Stanal does not sell this system.

Safety First

The lift and hoist methods discussed here are representative of what is now in use in the field. None of these items is foolproof. Accidents occur mostly because we are using these devices for purposes other

Figure 12-9 Lodestar chain motor
(Photo by CM Hoists Inc.)

**Figure 12-10 Flown
lighting system with
Lodestar motors**
(Photo by Sundance
Lighting Corp.)

than what they were designed for by the manufacturer. Safety should be the watchword for anyone using these methods. Consult the manufacturers. Most of them have become interested enough in the concert use of their products that they can offer suggestions on proper use. They want to protect you, and themselves from lawsuits.

Use a skilled rigger whenever hoists are involved to protect the building as well as the audience and the production company. In the final analysis, the safety of the performers and the audience should be our prime concern.

Figure 12-11 Chain motors with truss bridle
(Photo by Sundance Lighting Corp.)

Figure 12-12   Stanal flying system
(Photo by James Moody, rig by Stanal Sound Inc.)

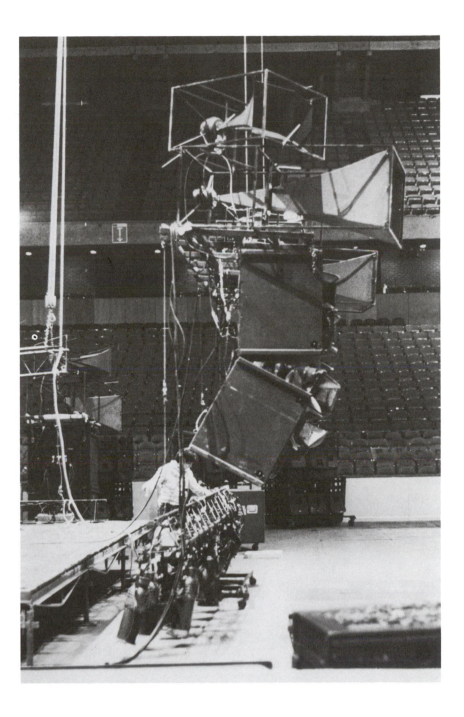

13

Moving Lights

The most radical innovation in lighting to come along in more than a decade has been remote-controlled (moving lights) lighting fixtures. The introduction of the Vari∗Lite® in September of 1981 by Showco Inc. in Dallas stunned the concert community, not only because of the innovative quality of the product, but because they were not a manufacturer looking to sell a product. In fact, as of this writing, they only lease the system.

How could the giants of the lighting world miss this idea? Actually remote control of fixtures has been around for a long time. From my trips throughout the world, specifically what I saw in Japan's NKH television studios and the BBC in London in the early 1970s, as well as reports out of Germany, I was aware of earlier usage of motorized lights. So what made this such an historic introduction? One explanation would be that the established lighting community saw it as a rock and roll effect *only* and didn't visualize the potential for it in theatre and television. This is not the first time the mainstream has been slower than rock and roll to pick up on a concept. (The list is long: PAR-64, trusses, multicable, portable packaging of dimming, et cetera.) Again tour lighting ingenuity had taken an existing technology and improved it tenfold.

The History of Moving Lights

While trying to put this new development into perspective I contacted some people who I hoped could fill in some of the historical background. Those conversations led me to Mr. Louis Erhart, a Yale graduate who had assisted the legendary Stanley McCandles from 1932 to 1934. He joined Century Lighting in 1937. In 1941 he helped establish the West Coast factory operations, retiring as vice president in 1972. I called him to find the historical facts on these foreign installations only to be told that American ingenuity had not been lacking. He produced a copy of a data sheet on a fixture Century marketed as the Featherlight. It was the outcome of a joint venture with Paramount Studios to develop a commercially saleable remote-controlled fixture.

The story of that fixture started at the end of 1949 when production began on Cecil B. DeMille's cinematic spectacular *The Greatest Show On Earth,* subsequently released in 1952. A unit was desired that could be mounted high in the Big Top without an operator. A Mr. Hissorich is credited by Erhart as the developer for Paramount. The joint venture ultimately produced a fully automated television studio in New York in the mid-fifties (NBC's Studio H). Unfortunately, some people thought that it meant the potential loss of jobs, so they reportedly did everything possible to sabotage the concept. It is too bad they did not realize that automation increases productivity, thereby increasing usable production time and the need for even more crew. That has been proven in Japan and London studios. Sadly, the Century/Paramount project was dropped shortly thereafter. They cited technical shortcomings in motor design, which could have been resolved if they had been willing to stick with the concept. Several of these units still exist and are in working order at Los Angeles Stage Lighting.

Current Development

The sudden rash of moving lights, motorized yokes, and computer-controlled fixtures, all names for the same general product, started with two nontraditional manufacturers: Showco Inc., a touring lighting and sound company from Dallas, followed a year later by Morpheus Lights Inc., a San Jose, California-based touring lighting company. The development of each product took similar paths but had interesting variations on the theme. Some dozen other companies are now in the field with many more offering new entries every month. Some are close copies, but to their credit, many have added features that have enriched this innovation. So the rock and roll computer light has obviously outlasted expectations that it was only a gimmick or short-lived effect light.

If the stigma of their being introduced by rock and roll companies caused the moving lights to initially be shunned by theatre and television, that prejudice seems to have been overcome. Television, especially, has taken to them, even awarding Emmys in 1983 and 1984 to two lighting directors (one twice) that used them extensively on specials. Corporate theatre is a big user of these units and Broadway has started to make them indispensable, starting with *Starlight Express*. Films, such as *Streets of Fire*, also used them in concert scenes.

Rock and roll has proven that each fixture, with its ability to be repositioned, change color, add patterns automatically, and even change focus on some units, is a very valuable tool. Even the least expensive units can give a big ballyhoo effect for your money.

Cost and Availability

Most of the small manufacturers of moving lights will not sell them but do lease full systems with an operator. As a purchased item the cost is unquestionably the highest of any lighting product we use. The lease expense is equally dear. With increasing competition, the costs may be lowered and there is every indication that fixtures more in line with theatre budgets will be made available. Already motorized yokes are appearing for use in television studios and other specialized areas.

So many new products are currently being introduced into this market that it is difficult to say what the next step will be. Certainly we have not seen the end to the creativity in this area.

Color Changers

Having multiple colors from one fixture is a very important part of these computer-controlled moving lights. We should point out that Showco led the field in this area also. Their approach was radically different. Instead of the boomerang type color changer, they used two dichroic filters arranged to rotate to produce a palette of some sixty usable colors.

The other new type of color changer came in the form of a scrolling device. The ColorMax was marketed in 1981 by The Great American Market (see Figure 13-1). It could hold twelve colors that could be remotely controlled in groups, accessible in three to five seconds depending on model size. It used the first microprocessor developed for

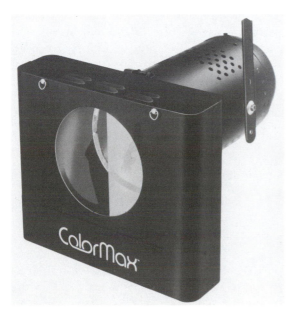

Figure 13-1 ColorMax, a scrolling color changer
(Photo by The Great American Market.)

this use. Virtually all the other moving lights use this or some type of similar system.

It is not possible to describe all the moving lights because the list would be out of date by the time you read this. What follows is a description of the two units that started it all and the recently introduced advanced versions of these lights.

Vari*Lite®

A separate company was set up to handle the manufacturing and marketing of this new product. The Vari*Lite® is described as "a self-contained computer-controlled lighting fixture." The unit consists of an upper box assembly that houses the lamp power supply, pan mechanism, and other electronics. The lamp housing, or head, contains the lamp, color mechanism, mechanical dimming system, and tilt control. The upper box is also where all the mounting hardware attaches. All of the fixtures are controlled by a multiplexed digital signal distribution system. This means that a single three-wire microphone cable from the computer provides all the control data for all fixtures.

The original Vari*Lite used a GE Marc 350-T16 lamp that could produce 140 footcandles at 40 feet with a color temperature of 5600 Kelvins. It takes two seconds to rotate the unit 180 degrees and the position is accurate to within one degree on either axis. It has a mechanical douser that goes from full on to full off in under one half second. The beam spread can be varied by choosing any of eight available aperture openings.

Probably the most unique feature, and until just recently, not copied, is the color system. The unit produces sixty colors using dichroic filter wheels rather than standard color media. It can change color in one tenth of a second. Besides the sixty preselected colors it is possible to dial in a mix of your own colors at the computer console.

The fixtures are all controlled via a custom computer console. The original console could store 250 cues from 96 fixtures. There is no tape or disc drive storage; the unit uses an integrated circuit storage. Cues can be written for each lamp or for groups of lamps and can be retrieved at will or in sequence. The board operator can manually manipulate any feature of any fixture during a cue.

But all this is *old* technology. Vari*Lite introduced their second generation light, which consists of two fixture types and a new console. The Vari*Lite 1™ is still on the road and will continue to be very useful for many years. However, the designers who have used this technology for a few years now are ready to move on, and this new generation has even more functions and design possibilities. Plate 2 shows the Genesis system of Vari*Lites.

The talk is now about "intelligent" lights. The Vari*Lite 2 (VL2)™ is a high-performance unit that can produce 1,000 footcandles at 20 feet using an HTI lamp (see Figure 13-2). Tilt has been increased to 270 degrees with panning ability of 360 degrees. The unit weighs 58 pounds and at 8½ inches by 17½ inches it is quite compact. The full unit, with control head and lamp, is 25¾ inches high. Now the unit has a precision iris and can be remotely focused to a hard or soft edge. It has an aspheric lens system. Reflecting advances in dichroic technology, the new unit has 120 colors that can be accessed in 0.12 seconds. Another unique feature is something called Vacu-Dep™ which allows the user to design custom patterns (very expensive). The unit holds nine standard or custom patterns on top of an internal "pattern/gobo" system.

Figure 13-2 Vari*Lite 2™
(Photo by Vari*Lite, Inc.)

The Vari*Lite 3 (VL3)™ is described as a "wash luminaire" by the manufacturer (see Figure 13-3). It uses a new 475 watt, 53 volt tungsten lamp at 3200 Kelvins that was custom designed for the unit. A beam spread ranging from an ACL-type beam to the field of a medium PAR-64 is possible with the unit. The light output is comparable to the VL2. While the VL2 has a mechanical douser, the VL3 has a built-in electronic dimmer. The size and weight of the two units are the same.

The color system for the VL3 uses their *Dichro-Tune*™ color tuning system. It is a fully tunable system that allows the designer to gradually dial through the spectrum to achieve the precise color desired.

All of these units are controlled by a custom computer called the Artisan™ (see Figure 13-4). A dramatically expanded version of the original design, it can control up to 1,000 fixtures and record 1,000 cues per fixture including color, intensity, beam size, edging, patterns, position, and motion. A bidirectional data link allows the console to have control over all functions and the status of each fixture. The location and status of each fixture is displayed on a touch-sensitive screen. Because the first-time user can become overwhelmed, eighty predesigned cues are stored for use as is or they can be modified. There is now disc storage of memory as a backup for the cues stored in the fixture.

Morpheus Lights

When Morpheus Lights entered the moving lights market shortly after Showco, they went a step further, making two different fixtures available. The short description that follows focuses on their first technology. They have also moved on to second-generation innovations.

The PanaSpot™ was a unit designed much like the Vari*Lite, with a single housing for all control functions and motors. The GE Marc 350-T16 was also their lamp of choice. A mechanical douser was used to dim the light, but they did have a fully functioning iris instead of the template idea Showco used. There is a slot for a mini-ellipse-sized pattern.

The beam size was altered by a magnifying iris. The beam varies from 2 degrees to 25 degrees. Color was via a boomerang setup of seven user-selected colors.

Figure 13-3 Vari*Lite 3™
(Photo by Vari*Lite, Inc.)

Figure 13-4 Artisan™ **console**
This console is custom designed to handle the full line of Vari*Lites. (Photo by Vari*Lite, Inc.)

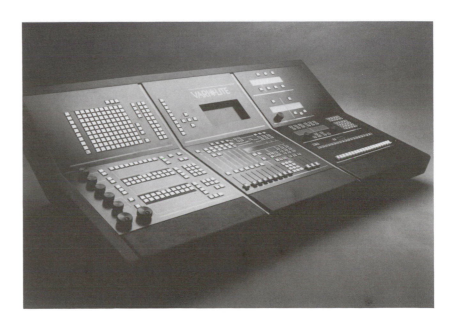

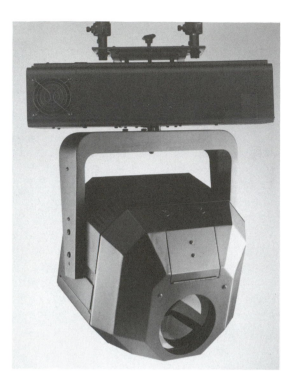

Figure 13-5 PC-Spot™
(Photo by Morpheus Lights Inc.)

The other fixture from Morpheus is the Panabeam™. While similar to the Panaspot in size and somewhat lighter, the light source is a standard PAR-64 lamp. Any beam width can be used including an ACL lamp. No dimmer is built in so provision needs to be made within the regular dimming system for intensity control. This unit uses a scrolling color changer with six colors and clear.

Morpheus took a different road by not designing a custom console. Rather they chose to use a stock Kliegl Bros. Performer 2 computer console, assigning each unit a position via the soft patch. It allowed for 125 lamps to be controlled and 225 cues to be stored. Their thinking was that because the console was mass manufactured by a mainstream lighting company it would control the moving lights plus any standard theatrical lights at the same time. Repair and replacement was easy via the Kliegl dealers around the world, so service was simplified.

The new family from Morpheus contains many advances. The PC-Spot™ (see Figure 13-5) has a new lamp source, an Osram 400 watt HTI. A significant factor is that the complete unit weighs 35 pounds, considerably lighter than the VL2 and VL3. The hanging center has also been reduced to 20 inches. The tilt coverage is not so good, coming in at 240 degrees. Proven zoom optics were once again able to provide a 2 to 25 degree beam spread.

This is the first unit other than Vari∗Lite to use a dichroic color system. However, it adds a ten-frame scrolling system to provide color correction, diffusion filters, or special colors the designer may require. Six pattern holders are built in and three are capable of rotating with programmable speed and direction, which makes for some very nice added movement in the light besides pan and tilt. Patterns are combined with the zoom lens to control the actual pattern size. Remote focus of beam from hard to soft is still possible.

The other new fixture is the PC-Beam™ (Figure 13-6). The lamp is a 1000 watt FEL and the unit has an internal dimmer. A parabolic reflector produces variable beam spread from ACL-type to wide floor PAR fields. Color is via an eleven-frame scrolling changer. Weight is a lean 20 pounds per unit, which includes dimmer, fixture head, and electronics. This unit also has preprogrammed looks built in; 100 are stored. They can be used as programmed or modified for user needs.

With this generation of lights a new custom computer console was developed. Called the Commander, it is smaller than the Vari∗Lite console and uses a detachable monitor for graphic display of cues and function of the fixtures. Because each fixture has its own on-board computer which, in this case, is an Intel 8088 processor with 64K RAM, the console does not need to contain all the functional information. This is another example of the "intelligent" lights that are the future in theatrical lighting.

Morpheus has packaged these lamps into a system that comes complete with trussing. A fold-up truss (Figure 13-7) contains twelve PC-Beams or PC-Spots or any combination of standard theatrical fixtures in each section. It should be obvious that if all moving lights are used, setup and focus time are dramatically reduced.

The Future of Moving Lights

There can be no question that the versatility of moving lights has application not only in concerts but in all forms of theatrical lighting. The costs are dropping. Recently I was quoted a price similar to a

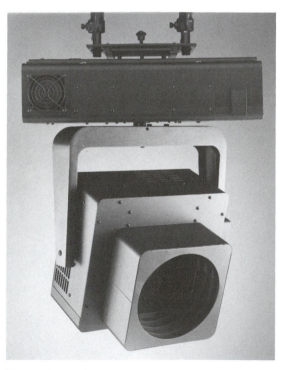

Figure 13-6 PC-Beam
(Photo by Morpheus Lights Inc.)

Figure 13-7 Morpheus truss with all computer-controlled lights
(Photo by Morpheus Lights Inc.)

standard 300-lamp system for 120 moving lights. There is no comparing the design flexibility that is possible over conventional lighting. The state-of-the-art quality and cutting edge of these developments are upon us. When you can focus, color, and even dim a source of such brightness and high color temperature remotely, the time has come to integrate our thinking.

Even more radical designs are being tested as you read this. The Satellite™ (see Figure 13-8), a unit from a new California company of the same name, is a 16 in sphere with an HTI lamp that can continuously rotate 360 degrees on both axes. The units have been field tested for some time and a public announcement will be made soon.

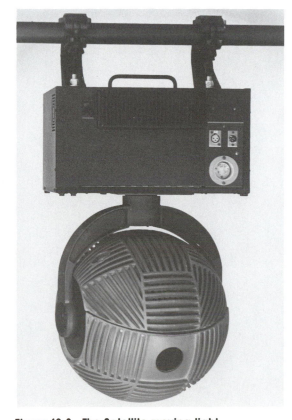

Figure 13-8 The Satellite moving light
(Photo by Satellite Illumination Inc.)

III

Practical Examples of Techniques

14

Concerts
In-the-Round

Parts I and II of this book covered the basic concepts of concert lighting. How you progress beyond the basics will be up to how you respond to circumstances and opportunities you find for yourself. No book can make you a designer. It can only explain current tools, procedures, and criteria. What makes a design is that inner voice that takes your accumulated knowledge and arranges it into something uniquely yours.

The next three chapters have been included to give examples of the lengths I have gone to achieve my designs. The projects described in these chapters have many elements that could be used for theatre productions also. The techniques used to mount and move the show from city to city were pure concert.

My choice of fixtures and how I used them may seem a great departure from what I have described as a typical concert. Don't let that throw you; all creativity is a departure from the norm. No matter what mix of tools you use, the object is to achieve a unique result.

With each project, I will explain the staging plans because, as I wrote in chapter 5, the concert lighting director is often responsible for all of the design elements.

What follows is an example of the integration of concert techniques, applicable to a theatre-in-the-round staging of a play or musical as well, that could be used in any place other than a specifically designed theatre space.

New Use of Arena Space

I received a call saying we needed to do an in-the-round setup for a John Denver concert. I had been doing his lighting for a couple of years, and we had always worked the show as an end-stage, as do the majority of concert artists. At the time I only knew of Frank Sinatra, Tony Orlando, and a few other Las Vegas–type acts who performed concerts in this manner. The problems were many, but the most difficult was that unlike these artists, John would not hear of his band being placed off the stage in a pit.

Under most circumstances the task would be to do a basic dinner theatre-in-the-round plot. Some nice full stage color washes and four or eight followspots, and I could take most of the day off. But not with John; he believed in giving his public a full production, including film and slides to augment his songs, and he had even used a full orchestra on several tours.

The problems of working this particular artist in this manner were

1. He was tied to a stand microphone, because he played acoustic guitar and sang.
2. He insisted on the band being on stage.

3. He did not want to obstruct the audience's view any more than absolutely necessary.
4. The large arena had to be transformed into an intimate club for the audience.
5. A fast setup was essential; John ran a tight touring schedule.

The general opinion was "Let's do these few shows and get back to reality," but I was intrigued by the challenge. It presented a perfect opportunity to do some crossover experimenting. I saw a fantastic opportunity to depart from the radically harsh concert look and present John not as a performer standing on stage with production behind him, but as a storyteller singing *with* his friends and fans. To create a feeling of intimacy in such a large space was a major goal. Part of what I felt I needed to accomplish this goal was to have control of the stage design.

Stage Design

The first problem in stage design to overcome was the stage itself. Dan Fiala of Concerts West, John's promoter on those concerts, went so far as to set up a stage and audience chairs in a shopping center parking lot to see how high the stage should be for optimum viewing. These same promoters were also handling the Sinatra shows and had experienced obstruction problems before. From Dan's investigation and ours, it was agreed that a low stage (two feet) was best. We found that if John was placed in the center with players around him, as you increase the height of the stage, you logarithmically increase the obstructed view for the audience.

The first leg of the tour had stages provided by each hall, which varied from two to four feet in height. The hall management fought us on the stage height almost everywhere we went, believing that crowd control would be a major problem. People jumping up on the stage could interrupt the show. We were fairly confident that John did not have the normal rock concert audience. Women did not try to tear his clothes off, and it was generally a family night, from mothers with babes in arms to grandparents. In all the shows we did in this manner (some 200), only three people got onto the stage. This was partly due to the audience makeup, and partly due to a good security and usher system. It was a system that used a very low-key approach designed to spot potential problems before they happened. Our feeling was that if you trust the audience to respect an artist enough to remain in their seats, the majority of them will do just that. Most problems occur when the artist disregards security and calls for the audience to disobey security.

As to the physical setup, John stood on an 18 in high by 6 ft round turntable in the center of the stage. The band on the first tour was relatively easy to work with; there were four musicians with minimal gear. Each man was placed at a corner, breaking the stage into pie wedges. An usher was seated, facing outward toward the audience, at each corner of the stage. We arranged for most seating plans to leave aisles directly in front of these points.

The next year, 1978, the problems were compounded when John brought in nine musicians, including three singers, and a baby grand piano. I spent several days constructing a model stage and photographing it from the audience point of view at each corner.

When I originally set up the stage with four musicians I used a Hopi ceremonial sand-painting design for the floor. I was interested in this art form, and it fit right in complete with an eagle feather pattern and the symbols of life from a Hopi ceremony.

With the expansion of the players, I knew the design would be lost, so I changed to a carpet to cover the stage. The carpet hid three items: the plywood stage flooring, the holes for the microphone cords, and the sound monitors so they could be connected beneath the stage, thus keeping the floor clear of cables. The carpet was a three-tone, golden brown, high-low cut. It held up very well through some 120 shows, even with all the holes cut into it. The redwood used around the turntable and the planters added to the earth-tone design, as did a dozen Boston ferns, which covered the backs of amplifiers. We rented the plants at first, then decided to purchase them because local florists were charging too much for a one-day rental. The plants lasted from three to five days while transporting them under these conditions, and the cost of purchase turned out to be half the rental cost. I made a lot of people happy because I would take the plants that were starting to wilt and distribute them among the facility staff and the stagehands. Why didn't we use artificial plants? They would last the whole tour. John wouldn't hear of it.

Lighting Design

There is no creative challenge in doing designs that are safe—ones that you are sure work. We should not discard our knowledge, but we cannot be slaves to it, either. When you, as a designer, are sure of your basic skills, then you are able to challenge yourself to find new ways to accomplish a task. This present project allowed me to do just that: take basic knowledge of theatre-in-the-round and combine it with my knowledge of concert colors and positions to achieve a new look for the show.

No Followspots!

To make the show intimate, I believed that followspots had to be eliminated. I felt strongly that I could light the stage in a very theatrical manner without them. Due to their distance from the stage, the beam's ambient light illuminated much of the audience. The angle, which varied radically from arena to arena, was the primary problem. In many venues, the followspots are no higher than twenty to thirty feet above the stage. This means that the first five to ten rows of people are blinded when looking directly at the artist. When other artists work in-the-round, they are constantly moving and therefore this problem is diminished somewhat. With an artist standing in one spot for two and a half hours, it would be unbearable to try and look into those beams for that long a period. The problem also extended to the fixed lighting. To achieve an intimate look, no single source of light can be overpowering compared to the rest of the light. A followspot is five times brighter than fixed illumination and I believed that would be too much for the soft look I wanted.

No matter how I reasoned, some management people insisted that followspots be used. The answer turned out to be simple, however. I told John to tell me if he felt the audience was being annoyed by the followspot light. It only took one song, after which he said over the sound system, "Jim, you were right, kill the spots," at which the audience applauded.

Technical Specifications

When I approach a design, I feel obligated to have a reasonable idea that the design can be accomplished. To that end, I, as a designer, must know the available equipment that can accomplish the task. If there is no available item that can do it, then I must design it myself or find someone who has the expertise to do it. This project created several new technical items. It also challenged me to find equipment that could fit into my plan and make the production workable. How this was done is covered below.

Mounting and Rigging

The problem of minimizing setup time was a constant concern when I was working out how the show could be hung. I wanted to have as few rigging points as possible, and because of the great cooperative effort between light and sound, the answer was already there. Stanal Sound's Ernie Zielinger had designed flying platforms for their sound system that formed a roof for the speaker cabinets to be hung underneath, rather than the conventional platform upon which to stack the cabinets (see Chapter 12). This allowed them to focus horns and cabinets at angles not possible from platforms normally used at that time. These grids were flown using 5/8 in aircraft cable with the motors mounted on top of the grid. The advantage in this rigging was portability and speed. The four sides required only two cables each instead of the common four-point system being used with chain hoists. These 5/8 in stranded steel cables were rated at 5,000 pounds each.

The cable drums they used were duo-sided so that both cables on the grid went through a pulley block to a single motorized drum mounted in the center of the grid. This ensured that the grid would go up preleveled.

The total load per side was certified at 3,000 pounds for both lights and sound. The cables, bridles, et cetera, required to hang the system are easily portable, and a second set was taken along so that the rigger could fly to the next arena the day before the show and do a prerig. When the usual chain-type hoists are used, their own bulk and weight rule this out, unless they are trucked.

Lighting Supports

The width of the sound grid was sufficient to allow a small triangular grid with lighting fixtures to be suspended from the onstage side without interfering with the sound system. Balance was maintained by moving the speakers forward. The lights were electrically isolated from the PA (public-address system) by wrapping tire innertubes around the grid between the cables attached to the light truss. Each of the four grids had individually controlled motors so that the grids could be flown to the height required for the best sound, even if that was at different levels in arenas that had high seats on the sides and low seats on the ends. A center grid was flown separately, either attached to the scoreboard in a gymnasium or with block and tackle.

The sound system must be positioned at a height equally splitting the seating, according to Ernie Zillinger, the show's audio engineer and mixer. This theory differs from other flying sound system designers who usually keep low (16 to 20 feet from a bottom speaker) no matter what the height of the building. I am not an audio engineer but I know our sound worked very well this way.

**Plate 1 Smoke high-
lights light beams**
(Photo by James
Moody.)

Plate 2 Genesis system of Vari∗Lites®
(Photo by Vari∗Lites, Inc.)

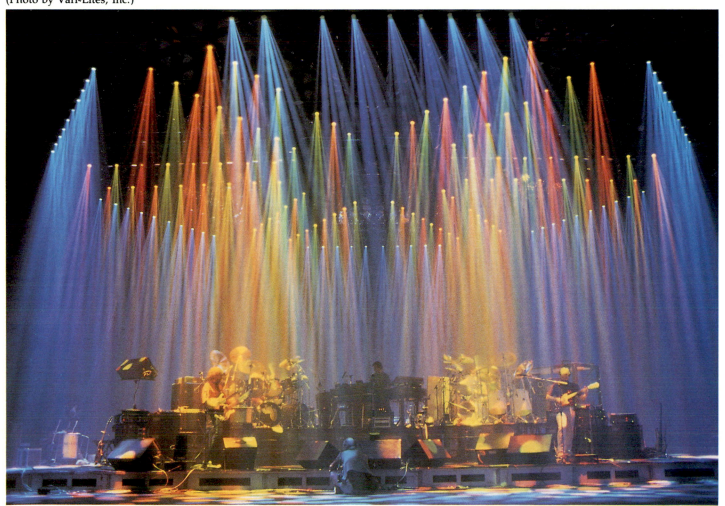

Plate 3   John Denver in-the-round show (Photo by James Moody.)

Plate 4   John Denver theatre design (Photo by James Moody.)

Plate 5   John Denver end-stage show
A 4 in × 5 in slide of the *Calypso* in background still reads clearly against the hot concert lighting. (Photo by James Moody.)

Plate 6   John Denver end-stage show, metal gobo patterns on the cyc
(Photo by James Moody.)

Plate 7   Music video shoot
(Photo by James Moody.)

Consequently, we had to be able to extend or reduce the cables on the light trusses to maintain a constant position no matter how high or low the building's seating. Lighting trim was best at 24 to 26 feet from the floor for this show. That allowed for the two-foot-high stage. That did make John's face light higher than a 45 degree angle, but more importantly, it kept the spill light on the stage and out of the audience.

Enclosed square trusses that travel with fixtures prehung and cabled, of which I am very fond, were not used in this case because of the extra weight, size, and light beam obstruction. Because the fixtures were not mounted in a box truss nor attached to bars in groups, each fixture's yoke was secured to the grid to ensure that no fixture could fall on the audience or a performer. Additionally, all color frames had safety wire cables attached to them. We tried to have people not sit directly under the grids. Management wanted a six foot walkway around the stage for traffic control and safety. But some halls pushed seats to within two feet of the stage.

Lighting Console

I had used computer consoles on the show since 1977. I started with a Siltron System 120 console and later moved to the Strand-Century Light Palette when it was introduced. I believe we got one of the very first ones produced. Because concerts do not get the weeks of rehearsals that theatre productions do to get each cue down perfectly, you must build the design as the songs are staged. The computer allowed me to start with the basic look and then build during the song and record what I saw. Because of the length of the show it actually took two 5½ in memory discs to hold all the cues that potentially could be used in the show. John did a stock opening set and closing group of songs, but the larger middle of the show would change from city to city. John was very sensitive to how an audience was reacting and would tailor the show to them. That meant that I could be called on to do any one of some fifty songs.

To make it easy to gain quick access to any song, I developed a plan that had each song starting at a multiple of ten, that is, song one started at 00, song two at 10, and song three at 20, et cetera. Because the Light Palette has the ability to add .1 to .9 between each cue I always had plenty of room for each song using this system.

I also programmed a series of basic opening looks into the additive submasters, so if unable to access a cue series quickly enough, I could bring up a submaster to cover until I could find the right cue sequence. It turned out that I did not have to resort to this backup very often. Even when an artist does not let you know which song will be next, you get a sixth sense for what will happen and, in this case, sometimes which guitar John picked up or how the band changed instruments would reveal the song to me.

Both consoles contained dual power supplies and battery backup so that instant changeover could be made if a power source were damaged in traveling or if a console lost power during the show.

Focusing

Because the triangular grids cannot be climbed on for focusing as most square trusses can, we carried two 24 ft Air Decks. They were also used for tying up the ropes once they had been hoisted up. Many halls did not have catwalks to allow the tails to be pulled up and snubbed off, so we tied the tail off and used duffle bags to put the loose rope into, then clipped the bags to the block to keep the look cleaner and make sure that the rope would not uncoil and come tumbling down during the show.

In focusing, the two Air Decks were used simultaneously on the arena floor next to the stage edge. The two men doing the focusing were pulled around the perimeter of the stage in opposite directions as they worked. The center grid was focused in a similar manner, but with one Air Deck on the stage before band equipment was in place.

Dimmer Racks

A 60 channel 3.6 kW Berkey-Colortran rack was constructed in the Sundance Lighting shop for the first tour (see Figure 14-1). Only the dimmers and their connector hardware were from Berkey-Colortran (now known as LEE Colortran). The case and all wiring for output panels, meters, et cetera were designed by Sundance specifically for road use. We loaded the 3.6 kW dimmers with 4 kW on a regular basis without any strain on the devices. The main breaker was a 600 amp motor-operated, three-phase unit with constant digital metering on all phases.

Later we built racks containing thirty 2 kW and three 6 kW Skirpan Lighting dimmers. Again, we designed and built the case, as well as all the interconnecting wiring, pin matrix patch, main and sub breaker panels, and output panels (slide patch).

Multicables

All the cabling from the grids to the dimming was 350 ft Cranetrol™ multiline cable. It took ten cables, two per grid plus two in the center, which contained nine 120 volt circuits terminated in Pyle-National's Star-Line series connectors for quick, secure attachment. The cable reels made it easy to store and move such long lengths (see Figure 14-1). We could fly the cables to the corner of the arena floor and not make the stage seem like a canopy bed with four cable posts around it. If the cables had been old single-circuit runs, we could not have handled the weight, nor would we have had the time to lay them all out each day and coil them each night.

Figure 14-1   Dimmer rack, with cable reels
The ten 350 ft multicables were stored and transported on the reel carts. (Photo by James Moody.)

Fixtures and Color Chart

Because John stood on a center platform on a 20 ft square stage, I used the fairly standard pie-wedge formula of in-the-round theatre lighting. I wanted to have very tight control of the light and contain it within the confines of the stage. To do this almost all 123 fixtures were ellipsoidals: a mixture of 6 × 9, 6 × 12, and 6 × 16, with only a few PAR-64's as down light (see Figure 14-2). Most of the 6 × 16's were the Colortran 20-degree models, which at the time gave the best lumen output. Certainly this fixture complement was unusual for touring. But the desire to control spill light was uppermost in my mind. As you can see from Plate 3, the lighting emphasized the use of people on stage as scenic elements, an essential ingredient in concert lighting even if there is scenery. Eight 1000 W Colortran 20-degree fixtures for (key) face light may seem like overkill, but I wanted two lamps on each corner (each on separate circuits) so that in the event of a burnout there would not be a dark side. Each truss therefore had two circuits (one warm and one cool color) of two lamps (one lamp at each end of every truss) to provide the key lighting.

In the center of the truss were 6 × 16 ellipsoidals that provided the full color washes (red, blue, blue/green, yellow, lavender, and white). Each color was broken into two circuits (North/South as one, and East/West as the second), and the shade of color changes from one circuit

Figure 14-2   John Denver In-the-round tour
(Photo by James Moody.)

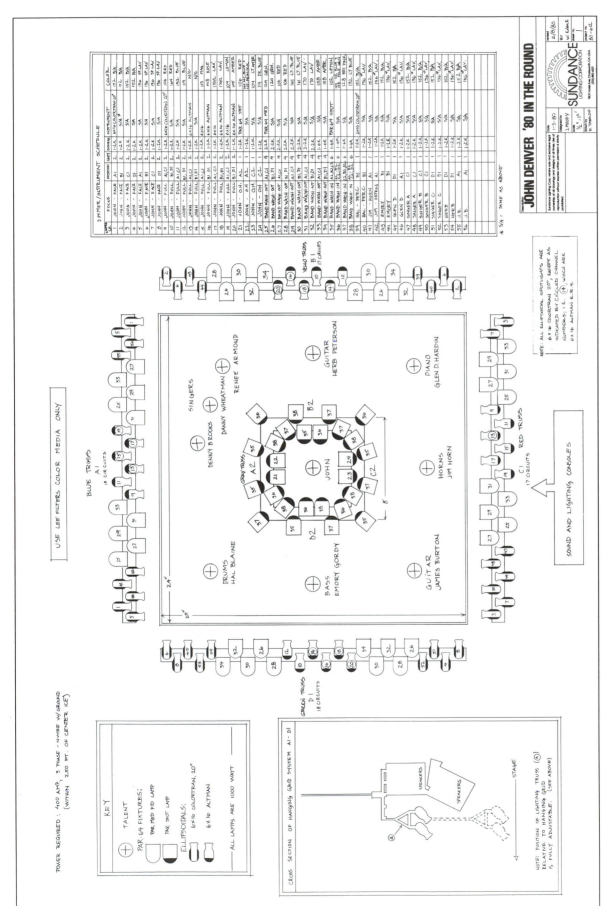

Figure 14-3 In-the-round light plot, John Denver show

to another, that is, a red #821 was on two sides and orange #819 on the opposite two sides. This gave a nice modeling to the body even if I only used the red full-body circuit from all four sides. Add to this four PAR-64 fixtures overhead as down light pools in red, blue, amber, and blue/green, to round out John's lighting. Figure 14-3 is the light plot, which shows rigging, circuit chart, and section view all on one page.

The band was illuminated with five color washes of eight PAR-64 fixtures equipped with medium flood lamps, two from each side. Again, the colors were divided and shaded from side to side as was done on John's washes. This procedure was repeated for the four color washes on the center truss. In addition, each band member had two key light specials (warm and cool) using a single Berkey 20-degree instrument.

Scheduling

John believed in a tight schedule. On the Spring 1978 tour, we played 57 shows in 61 days in 54 cities. That's a heavy schedule for any artist, especially when he did close to a 2½ hour show each night and frequently did second shows at midnight.

Critical Acclaim

The results of this design were displayed for over 2½ million people to see. Some 350 shows across the United States have been viewed and praised by audience and press alike. *Performance* magazine said of the Chicago Stadium show, ". . . the staging was a masterpiece of large concert production" (May 20, 1976). *San Francisco Examiner*'s Philip Elwood saw what I had tried to create when he wrote, "During the entire performance . . . the lighting was perfect, spotting soloists on schedule and making the small stage seem like a cozy general store where all the gang gathers 'round for a song fest" (May 12, 1978).

The United States Institute for Theatre Technology (USITT) presents a Juried Scenography Exposition. In 1982 this design was one of only five designs exhibited. It was the first time a concert design had been so honored.

15

Designing to Fit the Space

I was fortunate to design for an artist who welcomed change and encouraged it in his show. This chapter gives an example of another type of staging for John Denver. We often alternated between the in-the-round show and an end-stage or proscenium show because of facility availability. *End-stage* denotes a portable stage placed in the same fashion as a *proscenium*, a traditional theatre stage, or in the case of an arena or coliseum where seats circle the stage area, placed so that the majority of the audience will view the show from in front. As the designer you will want to be sure to find out if seats are to be sold behind the stage (usually with tickets stamped "Restricted Viewing"). You cannot use a backdrop if those seats are to be sold. You might want to negotiate to keep the backdrop and convince the promoter beforehand not to sell those tickets.

In some cases with the end-stage a full orchestra was added and I had to be prepared for them whenever they were used. However, as was true with the in-the-round shows, the foremost factor to John was the audience's unobstructed view. There was no scenery for the in-the-round design, but when I deal with an end-stage or proscenium show where there is no hard proscenium, the design concept grows. The majority of the audience is much farther away from the stage than they are for an in-the-round show.

In any concert design I want to create an environment for the artist that can make a statement about the music. If the group is Queen, U2, Duran Duran, or Kiss, you give them flash and nonstop color changes, like the pace of the music. If it is Dolly Parton, you drench her in beautiful colors to make the stage reflect her bigger-than-life image.

So, for "country boy" John Denver, I not only saw a man who had a very definite public image, but I heard in the lyrics of his music a very definite statement about family life and our use and abuse of Planet Earth. The public's concept of John is that of a lover of nature and the wilderness. The Rocky Mountains and the sea became an important part of the visual image he wanted to convey to the audience. My desire was to do a design that placed him in a scenic as well as lighting environment that matched his lyrics. The peace and beauty of lyrics such as

> You fill up my senses
> Like a night in a forest
> Like a storm in the desert
> Like a sleepy blue ocean*

had to be imparted to the audience visually.

*From "Annie's Song," lyrics by John Denver. Copyright 1974 Cherry Lane Music Co.

Stage Design

Before I could design the lighting I had to visualize the setting. Once I had an image I liked there was no longer a need for a scenic artist to redraw what I saw. So I also became the set designer. This is not unusual in concert touring. Unless the artist has the need for a very elaborate set, it is usually left to the lighting designer to at least specify the size and color of risers, backdrop, carpet, et cetera. This is 180 degrees from the usual Broadway method that has the scenic artists doubling in lighting so that they can ensure that the end product will have the color and shading they had envisioned in their scenic design.

This set for John Denver had as its focal point four 8 ft × 4 ft leaded, stained-glass windows, each depicting a season of the year and framed in redwood. We found a stained glass company to do the actual graphic design and they built the leaded-glass windows. I added a scenic element of redwood trim and support structure around them once they were finished. A rolling cart, well padded and with a protective cover, was specifically designed to minimize damage. To the great credit of this case and exceptionally fine leaded-glass work, the windows traveled all over the world without a major mishap.

The stained-glass windows were free standing to make it easy to adjust their position to match the physical restrictions imposed by the various stage sizes. If an orchestra was used, they were placed to the side so the full orchestra could be seen. Both floor stand supports and hanging hardware were carried to cover any situation we encountered.

The risers for the drummer and other musicians were also faced with redwood and carpeted on top. Planters with live green plants rounded out the set, which was all placed on a tri-colored (browns to red) carpet. Carpeting had worked so well for the in-the-round tour in tying the set together that I kept the element. Even the sound monitor system was carpeted to make it less conspicuous. John, however, stood on an authentic Navajo Indian rug, another subtle statement tied to the in-the-round tour motif.

Rather than the windows being the surface for changing images, I chose to complete the stage design with a full cyclorama (a curved curtain background). It served as a surface for film, slides using the Scene Machine, patterns, and color washes. The combined effect of the stained-glass window and cyclorama gave the audience the feeling that they were viewing the show from inside John's living room. Occasionally a stagehouse allowed for additional scenic elements such as this ceiling piece (see Plate 4). These elements were the mind's eye to his lyrics being suggested in images and light.

Lighting Design

Now that the set was complete, I had my surfaces to light, whether cyc or performer. In concert lighting our most important and one of the most useful canvases is the performing group. The first thing I plan in a design are cyclorama colors and the back light for the star and band. After I am satisfied that I have the colors and circuit control I need, then I work forward. The reason the cyclorama and the back light are viewed at the same time is that there must be a contrast between the colors so the back light does not get lost in the cyclorama.

It may seem odd at first that I start lighting from the rear of the stage. Theatre-trained designers usually start from the front of the house and work toward the rear of the stage. But I got involved in film production and learned that most directors of photography and gaffers (lighting electricians) proceed from back to front when lighting film. I find the technique similar to laying on the broad strokes for an oil painting. In film the canvas begins with the set (the walls of a room, et cetera) followed by general illumination that creates the mood of the scene. Only after that are the front light sources put in. If you pay attention to the layering, that portion of your audience sitting one hundred feet away will see the difference.

It must also be realized that in concert lighting the key light is known—the followspot—so less attention needs to be paid to it. It is the most common factor. I say this with reservation, but I have expanded on my feelings throughout this book.

My choices of colors are largely in the primary and secondary range. I am not prone to mixing gels in a single light source. When I apprenticed with Jules Fisher, the innovative Broadway designer of such shows as *Hair, Jesus Christ Superstar, Lenny, Dancin'*, and many more, one part of his design criteria especially impressed me and has had a great impact on my designing. He felt that with primary and secondary colors he could achieve almost any color during the cue-to-cue rehearsal. When you finally sit in the theatre with the director and he says he would like the set a little warmer, or he wants a magenta feeling, if you have picked colors that do not mix well, you are forced to re-gel and delay the rehearsal. Jules, on the other hand, had his palette in the air ready to mix instantly to his or the director's pleasure. And with time being money on Broadway, this technique helps to make him cost-effective to the producer. This is a simplification of his principle, but when I went into concert lighting I found this concept served me well. With minimal rehearsal time I would use my blue or red back wash or combine them to create magenta or lavender, depending on how I varied the level between the circuits. This saved me lamps and dimmers, both of which were at a premium in the early days of touring.

After the back light, I looked to the band side light, then John's side light. Next, pools and specials for band members, et cetera, were added. Just as with the in-the-round design, the contrast between colors on different levels or sections of the stage is very important in creating depth of field. Next the patterns for the cyclorama, floor, and set were determined. Lastly, like film lighting, the key-light colors were chosen.

However, in this design I had decided to eliminate all followspots since John does not move around the stage. I chose to use 6 × 16 ellipsoidals placed on a truss out in front of the stage in a fixed key-light position. My system provided for warm and cool cross-key face light plus a no-color circuit positioned straight on. Two lamps per circuit were used to increase intensity and as a safety against a lamp burn-out during the performance. After these straight-on key lights were focused at a waist-to-head shuttering, I put in several front-crossing full-body circuits of pastel colors.

I was shown films that had been produced especially for the show as backgrounds for several songs. One was shot on the ocean research ship *Calypso*, another in the Rocky Mountains, and yet another in black and white depicting a farmer in a field with a horse-drawn plow. My personal favorite was a beautiful film of soaring bald eagles. I had to be a part of the decision as to what order the songs were sung in because the films, slides, and other effects needed to be arranged throughout the set for maximum impact.

Because the films were done well in advance I knew which songs needed other visuals. So I started finding existing slides, or in some cases, I had them shot specifically for the tour. The slides were processed and mounted for use in the Scene Machine (Figure 15-1), a Japanese device. The 4 in × 5 in slides are great for realistic images such as the snow-capped peaks of the Rockies or the ship on which John wrote the song "Calypso" (see Plate 5). Besides holding film slides, the Scene Machine is excellent for hand-painted images. I also used another machine for moving clouds on this tour.

Lastly, I added metal gobo patterns for several songs that needed a more abstract image. John did sing frequently about trees and forests. I used some stock patterns but modified designs of the realistic stars and a reverse pattern of a stock leaf pattern done by the Great American Market. I have never understood why most people use a leaf pattern that makes the leaves the shadow area and the sky between the leaves the light transmission part of the pattern. Most tree patterns are done the same way. My design centered around first washing the cyc with the base color, that is, the blue sky, then laying over the white stars, the green leaves, or the green and brown trees. This follows the painter's broad stroke and finishing detail method. Andrea Tawil of The Great American Market and I created new designs of aspen trees and hanging leaves with branches, as well as an eagle (Plate 6).

The overall effect was to create an environment that reflected the music and gave a large enough image (cyc with film or slide or pattern) to be seen by the person in the last row. The audience needs a picture to view even if they cannot see the expression on the singer's face from 150 feet away (let alone be absolutely sure who it is standing on the stage).

This design is not a light show. This artist's music is unique, filled with beauty, not pulse-pounding power. But in either case I am still guided by the principle that good lighting is judged by where you do not put the light more than by where you do. It is very true that there are few cases in concert lighting that adhere to this philosophy. I firmly believe in it and fight to keep the quantity of fixtures down, not look for ways to increase the size of the system. See Figures 15-2 and 15-3 for the light and circuit plots.

Giving all your looks away in the first song by flashing through every preset you have is not design, and I take a strong stand against the flashing light school of nondesign. There is a difference between creating movement and excitement and just flashing lights to the beat of the music.

Before anyone says that my theories only work for John Denver or other MOR (middle-of-the-road) music acts, take a close look at the excellent recent designs for David Bowie and Whitney Houston (Allen Branton), Lionel Richie and Madonna (Peter Morse), John Cougar Mellencamp and Bruce Springsteen (Jeff Ravitz), Peter Gabriel (Jonathan Smeeton), Alice Cooper (Joe Gannon and Ed Geil), Genesis (Alan Owen), Bob Dylan/Tom Petty & The Heartbreakers (Stephen Bickford), Neil Diamond (Marilyn Lowey), or Jackson Browne (Marc Brickman), to name a few. The Rolling Stones 1982 tour, also designed by Allen Branton, showed great thoughtfulness of design.

Figure 15-1 Scene Machine
(Photo by The Great American Market.)

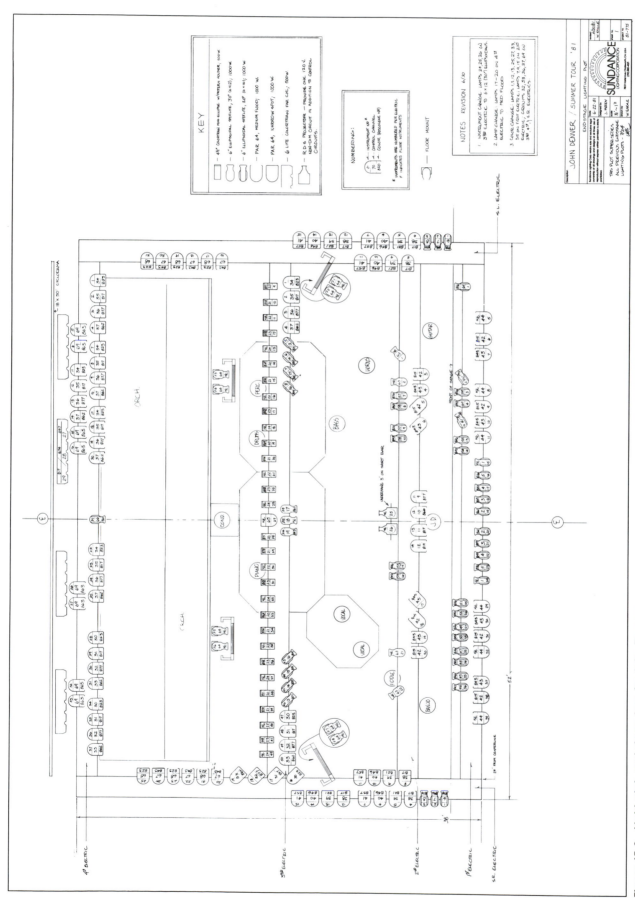

Figure 15-2 Light plot, end-stage tour for John Denver

DIMMER W/ LOAD	CONTROL CHANNEL	INSTRUMENTS TYPE	AMT	NUMBERS	LOCATION	DESCRIPTION	FOCUS	COLOR DESCRIPTION	NO/	NOTES
2K	1	6x16ERS	2	12,23	1ST ELECTRIC	KEY	JOHN	N/C	—	
2K	2			10,18				LT. BLUE	861	
2K	3			15,20				STRAW	809	
2K	4			14,21				SP. LAV.	842	
2K	5			17,19				FL. PINK	826	
2K	6	6x16ERS	2	1/1	S.R./S.L. ELECTRIC	SIDELITE		MAGENTA	838	
2K	7			2/2				DK. LAV	840	
2K	8			3/3				MED. GREEN	874	
1K	9	PAR M.F.	1	11	2ND ELECTRIC	DOWN POOL		BLUE/GREEN	877	
1K	10			13				URBAN BLUE	866	
1K	11			10				DK. AMBER	817	
1K	12			18				LT. RED	819	
2K	13	6x16ERS	2	5,47	3RD ELECTRIC	CROSS. BACKLITE		LT. GREEN	871	
2K	14			6,48				LT. MAGENTA	837	
2K	15			7,45				RED	821	
2K	16			8,46				STRAW	809	
1K	17	6x12ERS	1	25	3RD ELECTRIC	BACKLITE		LT. BLUE	861	
1K	18	6x9ERS		27				N/C	—	
1K	19	6x12ERS		28				MED. LAV.	843	
3K	20	6x9ERS	6	10,16,17,23,35,41	3RD ELECTRIC	PATTERNS	CYC	ASST. GREENS	871/874	PALM TREES
3K	21			12,18,19,30,37,42				MED. GREEN	874	ASPEN TREES
3K	22			11,20,24,34,36,40				N/C		REALISTIC STARS
3K	23			13,15,21,29,33,38				LT. GREEN BLUE	858	TREE TOPS
3K	24			14,22,31,32,39,43				N/C		CLOUDS
1K	25	RDS. PROJ.	1	14	2ND ELECTRIC	5 - 4x5 SLIDES		N/C		W/EDM DISC MACH. AND TP-5 TURRET
1K	26		1	15		5 - 4x5 SLIDES		N/C		PLATE - USE 0L·4 LENS
4K	27	CYC STRIP	4	—	U.S. FLOOR	CYC. LITE		MED. BLUE	857	
4K	28			—				MED. GREEN	874	
4K	29			—				DK. AMBER	817	
4K	30	PAR. N.S	4	50/31,35,39	3RD/4TH ELECTRIC	BACK WASH	VOCALS	DK. RED	823	
4K	31			49/30,34,38				DK. AMBER	817/2	DOUBLE COLOR.
4K	32			52/33,37,41				BLUE-GREEN	877	
4K	33			51/32,36,40				BLUE	862	MAY USE No 857 IF COLOR NOT AVAIL.
6K	34	PAR N.S/M.F	6	2/4,8,12,18,25	3RD/4TH ELECTRIC	BACK WASH	BAND	DK. RED	823	
6K	35			1/3,7,11,17,24				DK. AMBER	817/2	DOUBLE COLOR.

DIMMER W/ LOAD	CONTROL CHANNEL	INSTRUMENTS TYPE	AMT	NUMBERS	LOCATION	DESCRIPTION	FOCUS	COLOR DESCRIPTION	NO/	NOTES
6K	36	PAR N.S/M.F	6	4/6,10,14,20,27	3RD/4TH ELECTRIC	BACKWASH	BAND	BLUE-GREEN	877	
6K	37			3/5,9,13,19,26				BLUE	862	MAY USE No 857 IF COLOR NOT AVAIL.
6K	38	PAR N.S/M.F	6	4,5,12/4,5,12	S.L./S.R. ELECTRIC	SIDEWASH	BAND/VOCALS	DK. AMBER/2	817	DBL. COLOR
6K	39			6,7,13/6,7,13				RED	821	
6K	40			8,9,14/8,9,14				DK. LAV	846	
6K	41			10,11,15/10,11,15				MED/BLUE	857	
8K	42	PAR M.F.	8	2,6,31,32/1,8,21,27	1ST/2ND ELECTRIC	FRONT WASH		MED. AMBER	815	
8K	43			3,8,29,33/2,9,20,28				MED. LAV	843	
6K	44	PAR M.F.	6	10,11,13,22,24,27	1ST ELECTRIC	ON STAGE WASH.	3/4 EXIT & RETURN	N/C		
1K	45	6x16 ERS	1	28	1ST ELECTRIC	SPECIAL	BANJO	FL. PINK	826	
1K	46		1	30				SP. LAV.	842	
1K	47		1	25			FIDDLE	FL. PINK	826	
1K	48		1	26				SP. LAV	842	
1K	49		1	24	2ND ELECTRIC		VOCAL	FL. PINK	826	
1K	50		1	25				SP. LAV	842	
1K	51		1	22			VOCAL	FL. PINK	826	
1K	52		1	23				SP. LAV	842	
1K	53		1	17			PIANO	FL. PINK	826	
1K	54		1	19				SP. LAV	842	
1K	55		1	10			BASS	FL. PINK	826	
1K	56		1	12				SP. LAV	842	
2K	57		2	4,6			DRUMS	FL. PINK	826	
2K	58		2	5,7				SP. LAV	842	
1K	59		1	4	1ST ELECTRIC		HORNS	FL. PINK	826	
1K	60		1	5				SP. LAV	842	
2K	61		2	1,9			GUITAR	FL. PINK	826	
1K	62		1	7				SP. LAV	842	
3K	63	6x9ERS 750 W	4	1,2,21,22	2ND ELECTRIC	FRONT WASH	STAINED GLASS WINDOWS	N/C		
6K	64	6x9ERS 750 W	8	F1 - F8	MOUNT ON FLOOR	BACK WASH		N/C		
1K	65	PAR SPOT	1	26	3RD ELECTRIC	COND. FRONT LITE	CONDUCTOR	N/C		
1K	66	6x16 ERS.	1		4TH ELECTRIC	COND. BACK LITE		LT. BLUE	861	
6K	67	PAR N.S/M.F	6		S.L./S.R. ELECTRIC	ORCH. SIDELITE	ORCHESTRA	LAV	842	
6K	68		6					RED	823	
8K	69	PAR M.F	8	1,2,15,16,22,23,28,29	4TH ELECTRIC	ORC. BACK LITE		DK. BLUE	863	

Figure 15-3 Circuit plot, end-stage tour for John Denver

16

Rigging with No Ceiling

Even after the touring design has been done, there is no guarantee that everything will just drop into place to suit your plans. For a very unusual rigging application we will look to the John Denver show at the Universal Amphitheatre in the fall of 1980.

Problems

This was the last summer that the Universal Amphitheatre would be an open-air facility. Plans had already been drawn for a beautiful, new, fully enclosed facility. This show was to be the final performance under the stars. John's doing this final show was fitting because he had opened the facility in 1973. I had been there as the in-house lighting director at that time. The problems I had in fitting our production on the stage were as follows:

1. We were outdoors, with no covering over the stage.
2. The stage was just deep enough to get the set and orchestra on it, but the front of the stage had been cut with a very large radius, which would not allow us to keep the normal relationship of musician placement.
3. There was no apparent way to build structures that would support the trusses without obstructing the extremely wide audience seating plan.

Because of John's having opened this facility, both staffs were nostalgic and wanted to go out in style. Therefore, I had to try to mount the full production, including slides and patterns, with the full orchestra added.

Sky Hook

The solution to the main problem of truss support came, as many good ideas do, as a joke. "Let's get a sky hook!" And that is exactly what we did (see Figure 16-1). I contracted a construction crane company that had experience working on movies and sports events. Often a camera will be put in a basket attached to a crane for a spectacularly high shot. I merely wanted them to suspend our lighting trusses instead.

The mathematics were fairly simple. The weight of the truss was 2,800 pounds, the lighting cables were 1,500 pounds, and fixtures came in at 3,000 pounds, for a total of 7,300 pounds. This was not a problem for a crane with the required reach. Luckily there was an access road directly behind the backstage wall and we were able to get it into this area to the rear of stage left.

The last factor to consider was possible slippage during the week's

Figure 16-1 Crane behind stage
(Photo by James Moody.)

time the truss would be in the air. A hydraulic crane with a 160 ft mast was used because it had very accurate gauges that read out angle and boom-arm extension. I insisted on a guarantee in writing that it would move no more than one inch in the full week.

Set and Schedule Modifications

All of the above was accomplished in one twelve-hour day, including staging and band setup. Winds at night in the open amphitheatre were a concern so I did not want to use a cloth cyclorama or projection screen. Rather I had 4 ft × 24 ft hardwall flats built, covered with muslin and painted white. To frame the stage picture I added black soft goods from the top of the 50 ft rear wall, approximately 50 ft wide, but I did not put any drapery legs in as they would just blow around in the wind. The movement of the black drapery was not a problem because the hardwall white cyclorama was in front (see Figure 16-2).

I wish at this point it could be said that everything went off without a hitch. The fact is that after attaching the truss to the crane cable and putting on the fixtures, I realized that the lighting design was not completely symmetrical. Also, all the lighting cables were led to one point of exit off the trusses, stage left, and that along with the imbalance of the fixture layout put a lot of weight on that corner, which made the truss lean. The only solution was to add a guy-line to the back wall for balance once the truss was in place. We added two more guy-lines off stage left and stage right to keep the truss from turning like a top while suspended from the single center hanging point.

The actual performances started the next day after sound checks. The longer than normal setup was scheduled to allow for the anticipated problems that could result from the unusual production technique we were attempting with the crane. A break was planned between

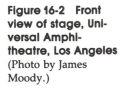

Figure 16-2 Front view of stage, Universal Amphitheatre, Los Angeles (Photo by James Moody.)

the expected completion of the setup and darkness. I wanted to be sure the setup would be complete before we lost the light of day. Then we could focus the lighting in the darkness. Because the system was almost half ellipsoidals and projections, shuttering and focus were more critical than with PAR-64's. I have focused PARs in bright sunlight many times but this was a very special show so we were allowed this luxury.

Rehearsal and the Shows

Day two started at one o'clock with a full orchestra rehearsal for the opening act: God himself, George Burns. Another fitting touch, as John had made his film debut with him in *Oh God.* Then came a rehearsal

Figure 16-3 Side view of stage
(Photo by James Moody.)

for John with the orchestra and we were off and running, almost. Actually, there was so much publicity generated by these being the final performances in the facility and the appearance of John and Mr. Burns together that the press ate up over an hour taking pictures and doing interviews.

Opening night went very well and we were ready to settle back for a nice run while everyone had a chance to go home to their own beds for a few days before hitting the road again. However, when we returned the next day, the crane operator assured us that the crane arm had not moved, but the crane body had! Apparently, the amphitheatre was built on the site of the trash dump for the movie studio and the cement slabs that were placed over this land fill had cracked at the front left leveling pad of the truck. The crane operator leveled the truck quite easily with the hydraulic leveling devices in the truck, but we were concerned that the ground would continue to give way. After consulting with the Universal Amphitheatre staff, it was decided that if the weight were distributed over a broader area the ground would hold. So, large 4 ft long, 2 ft × 2 ft wooden beams were placed under the leveling pads. That accomplished the weight distribution needed and kept the crane from sinking further.

The effect for the audience was unique. Many patrons had attended concerts in the facility for years, and such a large concert lighting rig had never been done since the stage had been reshaped a few years prior. The photograph looking across the front of the stage (Figure 16-3) shows how far the truss had to extend past the lip of the stage; if floor supports had been used, many seats would have been obstructed.

Although cranes have been used for outdoor festivals to hold up roofs, I believe this was a unique application of the sky hook, and it shows how a tricky problem can be solved when you are not afraid to try something out of the ordinary. When faced with a seemingly unanswerable question, look for an unexpected answer. Nothing is gained by saying, "It can't be done."

IV

Cross-Media Use of Techniques

17

Established Fields Benefit from New Developments

The previous chapters portrayed the birth of a new lighting media and the unique ideas developed to meet the lighting designer's needs. In the remaining chapters I will present examples showing how I have used these techniques and equipment in concerts as well as how they have been applied to theatre, corporate presentations, and video projects. First let us review some of the important elements in concert lighting and how they can be useful in other media.

Lighting Structures

The development of portable lighting structures is certainly concert lighting's single greatest achievement, already extended to all media. This was an area virtually unknown by theatre, film, or television. These highly portable structures have become the basis for development of new television studio lighting support systems at several new facilities. The labor coefficient cannot be ignored. The research and hundreds of thousands of hours of in-use applications in concerts must be taken advantage of by facility planners in the future.

The true advantage is not in classic design or in the engineering skill, but comes in the form of a practical structural concept that has been time tested. The size or weight-bearing factors will change with application, but the utilitarian construction of the devices must be studied further if full benefit is to be gained from what concert people have started.

Theatre Use

The use of trusses to house fixtures in trouping situations has moved to Broadway. Side-lighting structures where the lamps, whether PAR-64's or ellipsoidals, can be mounted, circuited, and stored all in one convenient case, are seeing very wide acceptance. Highly usable in this manner for dance troupes who want a convenient, quick setup system that is easy to operate, the truss adapts from its rear concert position to a side light (tree) position.

Other trusses can provide a multiple-use structure to handle scenery support, drapery track, or even as the skeleton of a portable proscenium arch.

Recently when Royce Hall at UCLA badly needed a second front-of-house lighting position, but could not use a permanent installation in the theatre, a truss was added. This was a practical solution to the problem. Two motorized winches raise or lower the truss to the floor when it is not in use, thereby retaining the architectural and acoustical integrity of the hall for functions not requiring heavy front illumination.

Television and Film Use

The value of trusses in television or film has even more possibilities than in theatre. Portable truss sections outfitted with 2 kW or 5 kW Fresnels that can be stored, then rolled into position for quick rigging on a sound stage or in a studio are a good example. On television locations, trusses prerigged with PAR-64's or Fresnels can provide audience lighting quickly and efficiently. The increased safety factor over a single pipe is a big plus.

PAR-64: Fixture for Any Media

The size and weight of the PAR-64 is a great advantage when fixtures must be quickly put into position on scaffolds or noncounterweighted pipes. The low maintenance factor is also appealing to the television stations, as is the comparatively low purchase price. Television has already made great use of the PAR-64. Audience lighting has virtually been given over to the PAR fixture. Other uses, such as musical numbers on shows from prime time to late night talk shows, now simulate the concert look with heavy color from PAR-64 fixtures. Television also gains greatly from the flexibility of the PAR-64 as a broad fill light on locations such as outdoor pageants, sports, and parades.

It should not take long to realize that open-faced sources have been a mainstay in theatre for years. Fixtures such as the scoop and beam projector have been in use for a long time. Because the PAR-64 is a self-contained lamp, reflector, and lens, theatre has questioned its controllability. But with four separate fixed beam spreads and several wattage sizes available to project from a broad field to a very narrow beam, the objection of poor control seems to lose steam. Theatre has traditionally been slow to accept change. So it is not surprising that the PAR-64 fixture is only now beginning to be accepted. Eventually, theatre will see the great opportunity it has to take advantage of this new idea developed elsewhere and make good use of it.

The use of the PAR-64 fixture has restrictions just as an ellipsoidal or a scoop does, but when used for punch lighting where no shuttering or dooring is desired, it will prove a very valuable tool. The fixture is available with accessories such as a snoot to help block and restrict some of the ambient (spill) light. Barndoors are also available for shaping of the light, but either one of these accessories is only minimally effective.

The greatest use of this light source is as back light, where reshaping of the beam is not as critical. Next in order of importance, it has great use as a special effect such as beams of light or intense color. When used as side light, the potential spill on set walls can be avoided if you are careful in positioning the units.

Dance companies could be the greatest benefactor of this new source. As they often use minimal sets and rely on light to convey mood and form, the PAR-64 is an ideal source for them. Generally, dance desires more color than legitimate theatre and thus will gain the benefit of the PAR's high light output to compensate for the heavy colors reducing lumen output.

Moving Lights

The remote-controlled fixture is currently the new bright star in concert lighting. Its impact is already being extensively felt, not only in concert lighting but in television, theatre, and film. Its ability to house different fixtures and lamp types (Mark 350, HTI, 3200 Kelvin) makes it even more appealing to these other medias.

I believe we are only seeing the tip of the iceberg. The second generation of these lights will open even more avenues of design possibilities. New developments are already under way to increase reliability and lower cost. Certainly we will see even greater use of these very flexible fixtures throughout the full spectrum of the theatrical lighting industry.

The initial high cost of renting these units has not helped their growth in theatre. However, as with any new product, the costs will drop when the market has been saturated. Low-cost "knock-offs" of the better units are already starting to appear on the market. In some cases the clones with fewer features may be just what dance and theatre need. The speed and large number of colors desired in concert lighting may not be necessary for most theatre work.

Multicables

When it comes to cables and the search for lightweight, flexible, cost-effective materials, again concert lighting did the field testing. While most installations are restricted to certain cable types by the electrical code, the concert people experimented, not always by the code, I concede. But what has evolved may very well rewrite the code. Currently, new materials for jacketing are being discussed for code consideration in the theatrical field. Again, the field studies were done, in effect, by the concert technicians. I am not condoning or encouraging the violation of electrical codes, I am just stating what we did to bring these materials out in the open for study and inspection.

The use of multicircuit cables for nonraceway theatres or portable facilities has many possibilities. They need not be restricted to portable truss situations only. The traditional taping together of single runs of cable to the end of a pipe, called a *hod*, is not as efficient as multicable on two counts: first, size, and second, weight. From a storage standpoint, a hod will take up more room and be heavier and harder to handle. It will add unnecessary pounds to an electrical pipe that is often too heavy anyway. Also, the hod does not coil well and is unruly to handle on the pipe.

Several years ago I took out a theatrical road show using hods and individually mounted fixtures. After we had mounted the show in a dozen locations around the country, we changed to new multicable (see chapter 20) and the time and weight savings were astounding.

Dimmers and Consoles

As for dimmers and consoles, it is true that it took the established manufacturers with their larger financial resources to develop computer consoles, but it was the push from concert lighting that made the dimmers smaller and more reliable. Packaging techniques for the road have been studied and incorporated into many of the newer models. If they can stand up to constant pounding on the road, they will surely handle normal installation situations in a relatively maintenance-free manner.

There is one idea in dimming that has not been greatly utilized by the major manufacturers and that is the 1 kW dimmer. Most of the research has been left up to small companies and designers. If a dimmer per circuit plan is of value to you, these companies warrant more investigation. Again, since their bread and butter has been made on the road, the reliability factors are in general very high and therefore maintenance should be at a minimum with these units.

While the major manufacturers were busy developing computers, the concert technician was looking to manual boards with more versatile functions. The old ten-scene preset board with its huge wing was not practical on the road. So development of pin and push-button matrixes was introduced in 1973 by Electronics Diversified. It is a very practical alternative to the computer both in size and cost and has validity in television as well as in theatre and dance. Here again, small companies did the major research and the benefits are now being seen in competitive prices to the consumer.

Summary

The major developments of concert lighting have been portable lighting structures or trusses, the PAR-64, moving lights, multicables, dimmers, and consoles. Now that these features of concert lighting have started appearing in other media, it is appropriate that I present some examples of how this has already been done. The following chapters describe actual projects that took advantage of this work.

18

Location Video Lighting

I got a chance to integrate my knowledge of concert lighting with video and theatre when I received a call from a Hollywood producer I had worked for on several projects in the past. He showed me plans for a set design with a false proscenium, two drapery tracks, and a cyclorama. Was it a game show? A musical/variety show? No, it was a television preacher's road show!

It seems Rex Humbard was no video flash in the pan. He was one of the first evangelists to see the value of the airwaves and had been appearing regularly on television for thirty-five years. This had led to a sprawling facility of two 60 ft × 80 ft sound stages, a 16-track recording studio, editing and distribution facilities, plus a 5,000 seat church equally well equipped for television.

All this was situated in Cuyahoga Falls, a suburb of Akron, Ohio. At the time, his show was aired for an hour over 550 television stations and over 1,000 radio stations around the world. They had been doing what they called "Rallies" (road shows), but this was to be his most extensive project to date away from his home facilities. Since this project, I have been involved with similar projects and they all demand topnotch equipment and network-level performance from their people and they are willing to pay for it.

This is only one example of what is being done on the road with video and concert media interchange. Showco Inc. sold Jimmy Swaggart a complete concert lighting system several years later. The project I discuss here could have been for any number of remote location video companies doing anything from a live game show to a variety show to the Miss America Pageant. Any traveling show can benefit from the concert lighting techniques to be described.

The program consisted of some nine musical numbers as well as Rex's sermon, all edited down to a one-hour program for distribution worldwide. The schedule called for playing in twenty-five cities in four weeks. Since then we have worked in several foreign countries. Overseas we used a modified schedule where we averaged one show every four days with location shooting in-between. In 1978 alone, we did some sixty programs, all on location (see Figure 18-1). The set, lighting, and staging were transported from city to city and set up in one day.

The Project at Hand

Naturally there are many limitations and problems to consider in a location video production. First, the nature of the performance involved:

1. a live show on location
2. minimal truck space
3. no rigging possible
4. quick setup and strike with a minimum crew and road staff

Figure 18-1 Location
television show, Rex
Humbard
(Photo by James Moody.)

These were just the physical problems. The artistic concerns included:

1. on-air lighting changes
2. isolated lighting areas
3. a cyclorama that was too close to the talent
4. a white set with thousands of 6S6 decorative lamps
5. a cast made up of Rex's entire family: wife, three sons, daughter, and two daughters-in-law and assorted grandchildren
6. new music production numbers every show

Through all this there was an answer—somewhere. We could handle the quick setup, no overhead support, minimum road staff and trucking problems; those, surprisingly, were the easy ones. Because of my concert background, I felt I could use these techniques to solve the problems. It still amazes me that even after some of the established television lighting directors see a touring concert with its sophisticated rigs and lighting control, they still discount its application in that media. I do not know if they cannot adjust or are just afraid to ask for assistance from some of the better concert lighting companies.

The Lighting Equipment

Using the financial base of Rex's organization and my experiences touring with bands, I was able to develop several specific equipment packaging ideas to fit the needs of the television production that had not been done before. Rex bought all the equipment from Sundance after we designed and fabricated it. The fixtures used were not the usual television fixtures and other ideas based on touring concerts were employed in 98 percent of the design.

Lighting Supports and Light Frame

Superlifts from Genie Industries were used for the floor support of all the lighting units. (These units were discussed in chapter 12.) I put two units in the audience area. Even on raked floors they worked because leveling jacks compensate for the slope. These were used to support the front or key light and for part of the audience lighting.

The structure I designed for the fixtures was a frame with Tee-nuts for attachment of the lamps. When traveling, the frame and lamps slide into a castered, open metal cage. All wiring was completed to a Pyle-National Star Line™ connector for fast electrical connection. Two men can set the entire unit in a matter of five minutes.

Dimmer

I knew that splitting the same power source between lighting, sound, and video would be a big question. At the time there were only two dimmers I would trust on the road: Skirpan and Colortran. I chose Colortran because at that time we used them on all our concert tours, and in five years had not had an RF (radio frequency) problem. Early portable dimmers were notorious for creating this type of transmitted interference, which sound equipment picked up as background noise. Also, at the time Colortran had service centers throughout the country and several overseas areas. Since this project was done, a lot of changes have taken place in the dimmer manufacturing field and there are now a number of dimmers that could meet these same road criteria. I want to explain how we arrived at our selection of equipment not as an endorsement but to show the process we used to devise these early portable systems.

The dimmer complement was forty-five 6 kW and three 12 kW "CRD" type plug-in modules, split into two racks for transport. Each rack had its own power distribution so they could be fed from different power sources or we had the ability to slave the panels together. Digital meters for all phases, amps and volts, were placed on each panel to ensure proper voltage and amperage. Actually, only the dimmer modules, shells, slide trays, and connectors were purchased from Colortran. The actual wiring and power distribution work was done by my company, Sundance Lighting Corp.

Console

The control console was built by Siltron Corporation, using the basic George Van Buren design for a 128 memory computer console (see Figure 18-2). At the time, three things were important:

1. the size—8 × 25 × 30 inches
2. delivery—under six weeks
3. the memory was contained on each 16 channel module, so expansion was done by adding 16 channel cards, with no need for disc drives

This early computer console was the smallest produced at the time. It was expandable by daisy-chaining plug-in modules. If a memory was lost you did not lose the whole system. A special road case was designed along with NiCad batteries to hold the memory between shows and as a safety factor in case power to the console should fail during the show. This small, and by today's standards, unsophisti-

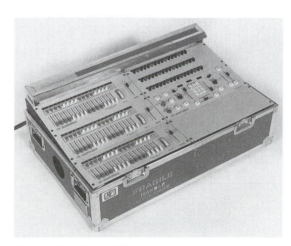

Figure 18-2 Early Siltron computer console
(Photo by Sundance Lighting Corp.)

cated console was large enough for our television production needs. It offered more presets, discrete level memory, and cross time or manual fades than a manual board.

Lights

The back lights were contained in a 12 ft, single-hung aluminum truss, which also supported the cyclorama. The side structures had to provide light in three directions, thus an unusual S-shaped truss was designed. They were to provide the downstage side light, audience front light, and upstage side light (behind the false proscenium). This design accomplished all three with the advantage of only one floor support required. That support was provided by a Genie Superlift, used in the audience area also.

The side structures could hold thirty PAR-64 fixtures each. The back truss could hold twelve PAR-64 fixtures. Each of the front Superlift cages could hold twenty PAR-64 fixtures. The total fixtures used were: eighty-five PAR-64's, six 6 × 12 ellipsoidals, six broads, two 12-light cyc strips, and eight 3-light cyc strips.

As is evident, I turned to concert lighting techniques and used mostly PAR-64's. The need was for the most efficient light for the least wattage. Also, the fact that there were no lenses to break or moving parts to jam was a prime factor in my consideration. Some Altman 6 × 12 ellipsoidals were used for patterns on the cyclorama and later six broads were added over the proscenium to fill in the upstage area. The cyc strips were already owned by Rex's production company so they were used to light from the deck (see Figure 18-3).

Packaging

The total lighting package consisted of the following:

2 cable boxes	2 Superlift cages
1 spare parts box	1 12 ft light truss
2 dimmer racks	2 S light trusses
1 #4/0 cable box	5 Superlifts
2 cyc strip boxes	2 Colortran followspots

Figure 18-3 Rex Humbard Singers
(Photo by James Moody.)

A total area of 30 ft × 7 ft of the trailer was used to pack the lighting and additional set and video equipment was packed on top of the lighting boxes to conserve space.

Technical Staff

Lighting Crew

When we first went out there were three lighting technicians: myself, a head electrician/board operator, and an assistant electrician. We later increased the crew by two who functioned as followspot operators during the show and electric crewmen during setup and strike. The need to have tight spot cues should be evident and this saved much valuable time in training new men each day. It also speeded up the load-in by having these extra hands to assist. The technicians were all from IATSE local 48 in Akron, Ohio. I cannot say enough about the quality of the union crew that traveled with the show. None of them were followers of the preacher, they were just doing a job, but with professionalism and a great attitude.

Total Technical Staff

Most of the crew was contracted for the road work only. A staff was maintained in Akron to run the facilities there. However, about thirteen of the road staff were full-time with the production company Rex operated called Cathedral Teleproductions. Eight stagehands were hired locally to assist in the setup and strike. The total staff consisted of:

1 Producer	2 Audio Technicians
1 Production Manager	1 Audio Engineer
2 Directors (one location, one live show)	1 House Sound Mixer
3 Video Engineers	1 Lighting Director
2 Video Maintenance	4 Lighting Technicians
2 Video Tape Operators	1 Makeup Artist
1 Video Switcher/Technical Director	1 Still Photographer
	6 Camerapersons

Packaging Update

Because there had never been such a complete video production undertaken by a religious organization before, there were bound to be changes. By keeping open to innovation, we discovered that the packaging was only at the starting point. Rex wanted to carry this elaborate show all over the world and the safe transport of the equipment was critical. Remember that this was in the days of 2-inch tape machines. The equipment was highly sensitive to movement in the first place. Someone proposed that we use the standard cargo containers, referred to as "Type A" by the airlines, to do more than just load boxes into. Why not build equipment into them so that they did not have to be fully unpacked? Just like some of our early concert lighting ideas, we all thought that the idea was crazy enough to work. We started calling them "the pods."

The first equipment to be built into such units included the video recorders, control room, and audio equipment. Two of these pods would fit end to end on a flatbed truck and could be off-loaded directly from the plane and go directly to the site. Once on site, all that needed to be done was to interconnect cables, attach power and they were ready to record the performance.

Then the use of the pod was expanded to the lighting. A plan was designed to mount the dimmer racks in a pod so that all interconnection to loads could be accomplished via Pyle-National Star-Line connectors mounted to the face of the pod. A second pod contained all the #4/0 cable and multicable to be run into the theatre, along with a shop space for spare parts, color media, lamps, and a repair area.

Two more pods were used to mount the mobile power plants, fully blimped. They purchased a 175 KVA Onan diesel generator for lighting, which produced 600 amps at 120/208 volts at 60 Hz, and a 45 KVA diesel generator for video and audio, which produced 200 amps, 240 volts at 60 Hz. This second unit also ran the air conditioners in the pods. Other pods handled the set, costumes, et cetera. With the addition of these power plants the production was *completely* self-contained as no other television production had ever been done, or has since been done in 1988. There are wonderful mobile video trucks but not a full show contained in a way similar to the ice shows, complete in containers.

All together the show took enough containers to fill a Boeing 747 cargo plane. But the same amount of equipment would have taken much more room if it had been stacked in boxes. Considering the time and labor savings, the project was a resounding success.

19

Music Performance on Tape and Film

Live rock concerts used to be just that—LIVE! If you missed them, your next opportunity to see the band came the next time they toured. Increasingly, however, rock concerts are being taped and filmed to be aired on cable or network television. Individual songs are used as promo spots or music videos such as those seen on MTV. Witness the recent outbreak of rock video venues: NBC's *Friday Night Videos, Hot Rocks* on the Playboy Channel, *Top of the Pops* which is shot on video in London and Los Angeles, syndicated shows like *MV3* and *Live From The Palace,* as well as full-length performances like *Diana Ross in Central Park* or *The Final Who Concert.* These shows alone add up to nearly thirty hours of programming per week. That does not include the full-time music cable channels such as MTV. That channel accounts for 60 to 70 percent of the music videos aired; recently a European version was launched.

Another link between video and concerts is the display of live pictures on huge screens while the concert is in progress. The use of the Ediophor or the PJ 50/55 large-screen video projection systems as an integrated part of the live performance is on the increase. Virtually all of the recent summer outdoor stadium tours used this means of allowing the fans 200 to 400 feet away from the stage to see the star "up close." Figure 19-1 shows the Ediophor set up in the back of the arena with the sound and lighting consoles. This Swiss-made unit is used all over the world and is considered the finest large-screen projection system built. Each comes with a price tag in excess of $200,000.

There are several companies that specialize in this type of video production. They work along with the tour lighting director to achieve an acceptable balance between what the designer wants to create for the live audience and what the camera needs to reproduce that image.

The lesson is simple: to be a successful concert lighting director and qualify for today's big tours, you had better know something about video lighting.

Film versus Video

Because of the increasing transfer of live performances to other mediums, debates rage between video and film makers over which medium most faithfully recreates the atmosphere of the live performance, and what are the technical and aesthetic merits of each medium. Film usually dominates for promo clips (commonly referred to as *music videos.*) Video is used more often for full-length performances, although some of the classic rock performances like the Band's *The Last Waltz* and Tina Turner's *Private Dancer* were recorded on film. The most recent Rolling Stones tour was both videotaped live and filmed by famed director Robert Altman for release as a full-length film, *Let's Spend the Night Together.*

Figure 19-1 Ediophor projectors
(Photo by World Stage Inc.)

Although I join a growing number of cinematographers who think a production should be recorded on the medium in which it will ultimately be viewed—the silver screen or the tube—there are technical considerations that often influence decisions from a practical perspective. The ease of editing, ability to capture the performance's ambience, and how much of the concert lighting can be readjusted are some of the considerations in choosing either film or video.

Film

The logistical problems of filming live concerts often determine the choice of media. Usually, five to nine cameras are set up to cover the action. Film cameras must be reloaded frequently, since 16mm and many smaller 35mm cameras can only take 400 ft loads, which translates to a maximum of 10 plus minutes of 24 fps (frames per second). Even the most popular major motion picture camera, the Mitchell BNC, can only accommodate magazines up to 1,200 feet and that is while the camera is mounted on a tripod. For hand-held work, the magazine length drops to 400 feet. There are cameras that can take 4,000 ft magazines, but they are not generally used in this type of photography. A concert does not stop for film reloads—therefore each camera is down for several minutes during critical performance time. There are seldom any retakes in live performances, so any missed action is lost forever.

Video

Video offers greater flexibility. A video director is in direct communication and visual contact with each cameraperson. Duplication of shots can be avoided. The director sees the "big picture," being able to view all the cameras at once. True, video assist is now common on motion picture cameras as is intercom communication to the director, but the picture quality does not give an accurate representation of the final image or focus because the video assist camera does not have the resolution of the film camera lens. Shots that the director counted on may turn out to be out of focus upon development of the film. But even more importantly, video allows *on-line* editing, electronic cutting of individual cameras directly to a master tape, which saves valuable post-production editing time and expense.

Depending on the budget and capability of the mobile video truck, the *program* (linecut) is double-recorded on two 1-inch tapes with a ¾-inch copy containing time codes for *off-line* editing use. The program is the real-time mixing of multiple video camera shots or taped feeds onto one master tape or onto a live line transmission. Off-line refers to the time the director can view the tapes without the heavy cost of an editor and the expensive equipment in an editing suite, another cost savings. One or two 1 in *iso* (isolated) tapes are recorded for cutting in the post-production editing session. Depending on the union situation, either the technical director, who also switches the linecut, or the assistant director will be responsible for switching cameras onto the iso feeds. Iso feeds are switched as straight cuts. No dissolves or fades are done on the iso tapes, whereas this is possible on the linecut, as are split screens and other electronic effects, which again saves time in editing.

Iso feeds are used for three reasons. First, they cover other action in case the on-line camera has technical problems or the cameraperson loses focus. Second, they allow the director to concentrate on the main action, although the director can ask for a particular shot to be isoed. Having that extra tape avoids having the director make quick judgments on unplanned shots. Third, the iso feeds can also be used to lay in audience shots or other cutaways for the final edit. They can cover a composition error that is not seen until editing or can be used as the second image in a split screen or other effects on the final cut.

Each director has his or her own way of switching isoes. Some keep the wide camera on iso throughout the performance and switch only close-ups to the program feed. Others have the AD (assistant director) keep an eye out and switch cameras into iso that are not the same as the on-the-line (program) camera.

The Debatable "Look"

Picture quality can be a debatable point when deciding on film or video. The "film look" is highly regarded as the equivalent to our concept of fantasy. Video is often talked about as too slick and real-life, too 5 o'clock news, for entertainment. You have to decide which image is right for your artistic goals.

There is now a movement toward melding the two media. Currently, there are two companies selling video cameras that produce a filmlike look. Panavision, a leader in 35mm motion picture cameras, now markets a video camera that uses their widely accepted film lens system and operation characteristics. In addition, its operating parameters offer a low light level feature that allows lighting to remain in the range of film production, a nominal 50 fc, instead of the 125 fc necessary for the average studio video camera to produce an f/2.8 iris setting.

Ikegami, a highly regarded video camera manufacturer, offers their EC 35 model with essentially the same filmlike look characteristics. However, they have stayed with the camera operator characteristics of the video camera, that is, no need for a focus puller person as with the 35mm film cameras. It should be noted that they will equip the camera for the inclusion of an assistant if desired by the cameraperson. The assistant usually manually or remotely changes the focus of the lens while the camera operator needs both hands to operate the gear head for pan and tilt on a normal motion picture camera.

Many people in video feel the EC 35 has the best electronics currently available. Its image enhancement circuitry uses a low noise device, resulting in a signal-to-noise ratio of over 57 dB rms (root mean square), and a new low capacitance diode-gun tube capable of producing an extremely sharp picture edge-to-edge with improved resolution and detail.

A cross-media trend is now developing in which performances are shot on film, then transferred to video tape for editing and viewing. There is a film format that has an aspect ratio of 1.33:1 for both 16mm and Academy 35mm, the same that is used for television, but wide-angle Cinemascope type lenses should not be used because too much of the image is lost when viewed on the home television receiver. It is hard for a cameraperson to keep in mind the Television Academy "Safe Area" when shooting and that is why you see boom mics in pictures on television series shot on film more often than when they were shot on tape. Here again, the director and technical director can *see* this happening whereas with film it will not be apparent until screening the next day when it is too late to go back and reshoot.

The Television Academy technical specifications state that as little as 35 percent of the image on the standard 35mm film will actually be received in the home via television. The "Safe Area" for video represents only about 70 percent of the total aperture area of the 1.33:1 aspect ratio of the film. So you can see why it is so easy for critical action to be cut off in film viewed on television.

Lighting Considerations

The inevitable conflict between the recorded media's lighting needs and the obligation to the live audience, who paid good money to see the concert, is always a tough fight. The normal concert lighting look has to be broadened for these other media. A compromise must be reached or the recorded product will suffer. Your job as lighting designer is to enhance the artist's image. If the tape or film is bad because you would not compromise your concert lighting, the artist is the loser.

When I first started doing rock video in 1972 as the lighting director for *Don Kirshner's Rock Concert* series, I had already logged six years of concert tour lighting. The concert lighting director is, in fact, the concert "director." We direct the audience where to look; we produce the visual picture.

In video or film, the concert lighting director becomes subservient to the director, who chooses the image, framing, and other shots to be recorded. Therefore, the concert lighting must be broadened to facilitate these needs, something the "live" show usually avoids. The video lighting director has to use a broader brushstroke when lighting. Plate 7 is a music video shoot. While I would not have lit the trusses as much on tour, here it works.

How can we best handle the live show so that video or film and live audiences are equally happy? Generally, I find that a concert does not need to be completely relit. Rather, *balance* is the key. The best video cameras have a contrast ratio of 32:1, while film has a much wider latitude (from 64:1 to 128:1). Video gives you five f-stops, as opposed to film, which gives you eight or more. As a result, video lighting ratios should not exceed 3:1 in the overall picture balance.

The 2:1 ratio is used to teach video lighting, but concert lighting can exceed this to help give the lighting that raw-edge quality.

What the Camera Sees

The best way to check how the camera will "see" the stage is to purchase a *contrast filter* for about twenty dollars. The Tiffen Company's model, which is widely used in film production, works well in video if you get it with a 2.0 neutral density filter. Hold the glass to your eye and watch a live concert to see how much of the detail will be missed by the video camera's reproduction system. This same procedure can be used with film once the film stock and its properties are determined. A different filter, with the appropriate higher contrast, will accomplish the same thing for film.

Another part of balance is color. The contrast filter will also show you how some color combinations are lost when recorded by the camera. Remember that the video camera has definite limitations, and one of these is *saturation*. In the three-tube Plumbicon or Saticon camera, a beam-splitter breaks the light received through the camera lens into red, blue, and green signals. The green channel sees more light in the visible spectrum. The red channel is almost half as sensitive and cannot produce enough beam voltage to eliminate residual image retention. That residual image, or comet-tailing, can generally be corrected by reducing the saturation of the color. *Lag* means the tube is not receiving enough light energy for it to provide a sufficient signal to the receiver. It gives you a ghostlike image and noise. Also, too much of any one color will cause tearing, which looks much like lag but is caused by color saturation.

Because of unequal color sensitivity and the reduced contrast properties of camera tubes, color shades will not always reproduce exactly as we see them. Therefore blue-green will come out green because the green tube is more sensitive. Oranges and some magentas will appear red and lavenders will turn blue on the television receiver. The video controller can influence this to some degree, but the cameras and which tubes are in them are the final controlling factors. You have to experiment to get an understanding of this problem.

Also, if you have film or slides in the live show they probably will not be bright enough to be recorded on the tape. Slides and film are best added electronically during post-production through effects processes. Remember, you are lighting for a broader view of the show and what the eye sees is not what the camera sees with its very limited contrast ratio. Lighting on scenic pieces generally must be increased in intensity if they are to "read."

There are ways of compensating for this in video through the efforts of the video controller. The controller watches and controls, among other things, the iris of the video camera. However, you should not count solely on this ability to boost the level electronically to make the light intensity acceptable.

Cutaways

The often repeated example of why cutaways are important is the five-minute guitar solo that goes over great with the live audience but is dull when transposed to film or tape. Television viewers have very short attention spans and must be kept interested with visual images that add to the enjoyment of the music. Extensive studies have been made on how often to change images if you want to keep the television viewer's attention. The director must use cutaways such as the live audience's reaction to the solo, or other band members' reactions to insert during the solo.

Key light is generally thought of as a film term. The key light is the primary source of illumination from the direction the camera views the scene most of the time. It is this source that very often has the lowest footcandle reading on the concert set. In television it will often be the brightest. In concert lighting, the back light is usually the brightest. Because followspots are usually the concert designer's only front light and because they cannot be trained on all the musicians all the time, it is unlikely that the camera will have enough illumination for other shots, or cutaways, on the drummer, keyboard player, or other individual band members.

The lighting problem on cutaways can be solved in three ways. First, add followspots. Not always an artistically justified solution, I know, but it will do when no additional fixed lighting can be added. Second, adding an even bank of front white light producing 125 fc will help. The video controller will be thrilled, but it takes away from the audience's interest in what is happening on stage. Third, try placing white light specials from the front on the drummer and keyboards and backup singers that can be dimmed up only as required. This will cause the least change in the live look while satisfying the video controller and the director's needs.

Audience Lighting

Audience lighting for taping or filming a live show is a must. Without it, the show might as well be shot on a sound stage where camera placement and lighting are optimal. What makes the performance *live* is the audience's reaction to the band. The producer/director needs that reaction on tape. A talk with the director will yield definite ideas on how to handle audience lighting. Generally, the discussion will focus on four questions:

1. Should the light be colored or white?
2. Is the audience light to be on all the time or just between songs?
3. Does the director want front, side, back light, or a combination on the audience?
4. How much time and money can be spent to mount additional lighting for the audience?

I do not like to light an audience from the onstage angle as it puts light in their eyes constantly and makes it difficult for them to concentrate on the stage performance. Back light is great for showing the size of the audience on a wide shot, and side light will pick up enough faces to satisfy most directors without annoying the whole crowd. You see many audiences lit with colored pools of front light, especially in the back of the auditorium or arena. It looks great on camera but is very distracting to the audience.

Just accept the fact that whatever you do it is a no-win situation and make the best of it. The audience will hate being distracted no matter what you do, so you might as well give the video or film what it needs with as much consideration to the audience as possible. I recommend side light as being best and only brought on between songs and during a few fast numbers. Remember the reactions can be cut in anywhere so the specific song being performed at that moment makes no difference as long as there is no reference to the stage in the shot.

Accommodation

Whatever the lighting director does to accommodate video or film on a live concert will have adverse effects on the artist and the road crew's normal operation of the tour, so try to understand the crew's problems and the pressures they are getting from the artist. If you go into this with an open mind and a willingness to cooperate, much can be accomplished and an exciting show can be recorded that satisfies everyone.

20

Corporate Theatre

About twelve years ago I was invited to see a show written and produced by Milliken Industries. The show was a full-scale professional Broadway musical presentation, designed to promote fabric names to clothing manufacturers while entertaining their clients for an hour or so. It seemed an unusual and novel idea. Early in 1977 I received a call to do a musical sponsored by the Amway Corporation, a Fortune 500 company dealing in home and personal care products sold through their own in-home marketing plan. Since that time I have designed other shows for them, as well as for many auto and motorcycle manufacturers and such unlikely corporations as the Bank of America, Avco Finance, Digital Computer Corporation, IBM, Avon, Coca-Cola, Sunkist, Federal Express, and Wendy's Hamburgers.

Why would they want a theatrical production? They all see this as a form of business meeting, pep rally, and entertainment package all rolled into one—sort of a sugar coating for the hard facts and figures they must deal with in business. Of equal importance to most, the show is a method of getting their employees or dealers motivated to sell more products in the coming year. In the case of Amway, a direct selling organization, they used a series of regional shows to talk to the local dealers and in turn, the local dealers used this as a way to bring new people in as distributors. As you might expect, these corporate shows involve special considerations for a concert lighting designer.

The Amway Show

The Amway example covered in this chapter, although not the biggest or the most recent, uses a lot of the mixed media that has already been discussed in this book. The original production did not use video projection, but in subsequent productions it was incorporated.

Amway is very vocal concerning American free enterprise and they decided that an original musical extolling the virtue of free enterprise would convey their message. They called it "Faces of Freedom" (see Figure 20-1). Their simple idea took a script, music, sets, lighting, director, choreographer, et cetera, and cost $1.4 million! That price did not include the cast either. (In 1982 I designed a onetime show for Cambridge Diet Plan International in Hawaii that cost over $2.8 million [see Figure 20-2]. To get an idea of the size of this production, the video projection in the figure is forty feet high.) So you can see that in all of live entertainment these productions spare no cost and are easily the most expensive productions short of Broadway.

Amway's initial project came out of a Bicentennial program they did in 1976. Its reception at their national sales meeting prompted them to think about a big traveling show the following year. The point was to produce a vehicle that would support a theme of American free enterprise and the pride of ownership of your own business, the foundation of the Amway sales program.

Figure 20-1 "Faces of Freedom" show, Amway
(Photo by James Moody.)

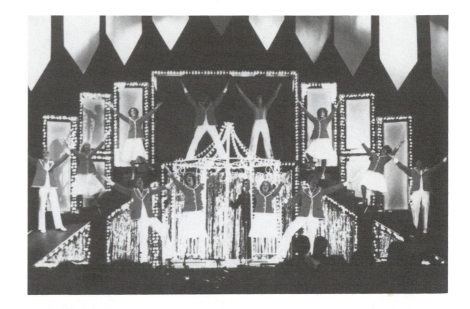

Figure 20-2 Cambridge Diet Plan International show, Hawaii
(Photo by James Moody.)

The script was written and rewritten several times. Original tunes were composed, and a couple of old standards were used for a barbershop quartet number. The cast was recruited and auditioned from the ranks of the Amway "family," the only nonprofessional element of the production. On their behalf, let me say that they were all seasoned performers who used the Amway plan to help them get by during their struggling days of acting, dancing, or singing, and found a real business to support their creative goals; call them semipros at the very least.

A one-day setup and show schedule was to be strictly followed. All of the technical services employed had to be geared to accomplish a very efficient operation.

The Lighting Design

At this point in time, 1977, concert techniques were really foreign to theatre, so I was in a unique position not only to design a legitimate lighting plot, but to use the techniques we were developing on our touring rock and roll concerts. As it turned out, however, the lighting budget was too low. Most of the production budget, as usual, went to set and the multimedia effects. Nevertheless, I felt comfortable with the design and the basic light plot. Since this was not a concert but a full musical production the design followed fairly well accepted practices with a few exceptions.

The front light could not be hung in a normal theatrical front-of-house position because of the short setup time, and because some of the halls did not have the physical location needed. The show did not play the usual road show houses. In fact, several shows took place in concert halls with no flies available at all: Constitution Hall in Washington, D.C., and the Los Angeles Sports Arena. Two house follow-spots were required for key light during dance and musical numbers in the usual manner. Two Genie Towers were placed just off the front of the proscenium as crossing front washes.

Three house pipes were required, but in halls with no flies, concert lighting trusses were rented locally to provide the positions. The two rear trees provided effect light on the screens seen above and behind the upper steps of the set (see Figure 20-3). The total lighting effect was accomplished with thirty-six channels of dimming and 106 fixtures.

Setup Time

The lighting equipment bid was awarded to one of the best theatrical rental firms. It took eight hours to unload, hang, and focus using the traditional theatrical methods of mounting each lamp separately, running hods of single cables, and putting together dimmer packs and connecting to a bull switch.

The production manager called me after about two months and complained about the amount of time it was taking to complete the lighting setup. I reminded him that in pre-production discussions I had said that if they would use the rock and roll type equipment and handling techniques, we would save several hours in setup time. It took another couple of months before he convinced the producer and Amway that the additional cost of the concert equipment would save time and that translated into reduced stagehand bills.

The change was made nine months into the run. Certainly that was enough time for the crew to get the setup down. The first show setup after the changeover, and I might add, with only one crewman knowing the changed equipment, was cut by two and one-half hours. Over the rest of the run the average time saved was four hours. Here again,

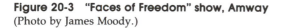

Figure 20-3 "Faces of Freedom" show, Amway (Photo by James Moody.)

as in the Rex Humbard television show (see chapter 18), the techniques of concert lighting made a difference.

What Was Changed?

The trees were replaced on Genie Towers with all fixtures attached, colored, and wired to one multiconnector. Add one multicable connection and air and the unit was in place. For the three overhead pipes, stock unistrut track was used with five fixtures mounted and cabled to a multiconnector mounted on the unistrut. A frame for roading these pipes was devised for storage and transportation. A single multicable was run to the end of the pipe and all five light bars were plugged into it. The savings here were in labor and time. One person can handle a #12/20 S.O. multicable; it takes two or three people to lug a nine-cable hod of #12/3 S.O. to a pipe and secure it in place.

The dimming equipment was also reworked into a road case and was prewired to help decrease the setup time. Meters on the main breaker panel allowed a quick visual check of amperage before turn on.

A consideration for the board operator was also included. The first system used a five-scene preset board. The second system used Electrosonic's Rockboard, which had a two-scene, three-subscene board with ten matrix presets and five push-button presets, thus giving the operator twelve more presets. This made for less complicated resetting of presets during the show.

All of these tricks were learned during the early years of concert touring. Again, the "bastard" media of rock and roll lighting had shown itself to be a proving ground for new ideas and techniques in another media.

The Second Year

The next year and for the next several years, the production for Amway Corp. was changed to a partial business meeting, a talk by the President or Chairman of the Board, and three or four musical sections. The shows were produced almost exclusively in arenas because of the size of the audience they were attracting. Then the complete lighting packaging took on a total concert look. As can be seen from the light plot in Figure 20-4, the arena show utilized a full concert flying system. It was designed for the higher illumination levels required for video. The side view shown in Figure 20-5 was needed to work out maximum truss height for best video lighting. If the truss angle was too high the close-ups would have eye socket shadows.

But another dimension had been added: a large video screen was introduced so that the 15,000 to 20,000 people could see the faces of the speakers as well as the performers (see Figure 20-6). Not only was there a need to dramatically light the singers, but enough illumination was necessary to get a good video image. A three-camera video package was added with an Ediophor projector to provide the large video magnification.

Because the production stage increased in size, not only had the amount of lighting grown to eighteen 6 × 16 ellipsoidals, eight 2 kW Fresnels, and 247 PAR-64's, but it was flown and all lamps were mounted prehung in trusses. The total load-in and setup, including staging, video, and flown sound system, was accomplished in six hours!

All cables were multiconnector nine-circuit cables. The dimming was

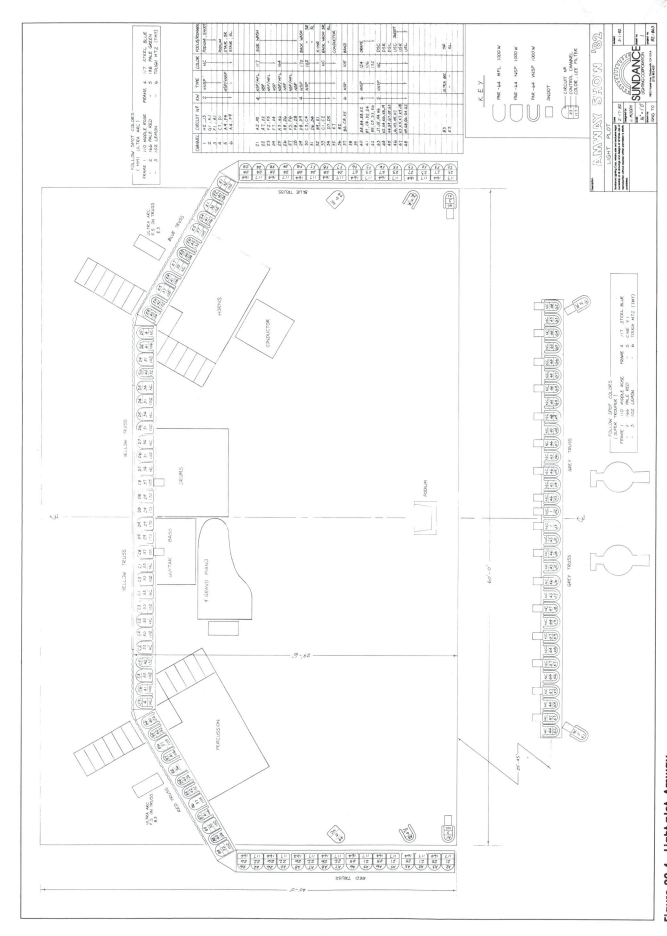

Figure 20-4   Light plot, Amway

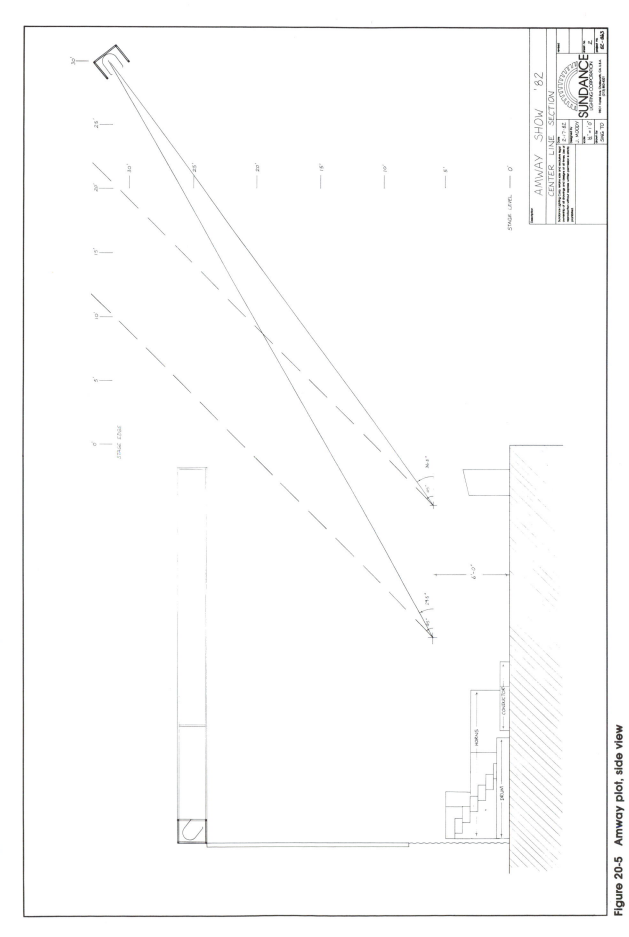

Figure 20-5 Amway plot, side view

Figure 20-6   Amway show with video projection
(Photo by James Moody.)

produced by Colortran 4 kW electronic dimmers and the same Electrosonics console used in the first year's show.

Of course it took a lot more trucking space to handle the set, screen, and video, but the fact is that the setup time actually decreased. Our packaging and handling did the trick (see Figure 20-7). Here we have the front view of the lighting rig that also supported a 35 ft drapery panel and a 24 ft × 32 ft rear projection screen.

A New Lighting Field?

Combining the video, rock and roll, and live musical lighting techniques in one production has become normal to me. It seems most of the corporate shows require all of these elements. Another example is the Cambridge Diet Plan International show in Hawaii, mentioned earlier, which had over 1,000 lamps all prehung in trusses and flown to the islands for the show. Load-in took only two days, including the on-site construction of a 120 ft by 30 ft screen for front projection of slides, film, and video. A 6 ft raised stage equaled the width of the screen and was 80 ft deep. It even had an ice rink built in it. They did all that for three days of speeches and shows for distributors of their diet products.

A motorcycle manufacturer put on a big show in Las Vegas that required lasers, pyrotechnics, projection video, moving lights, special effects, and a lot of smoke (see Figure 20-8). The smoke found good

Figure 20-7   Amway staging
(Photo by James Moody.)

Figure 20-8   Yamaha motorcycle show, Las Vegas
(Photo by James Moody.)

use in dramatically revealing the motorcycles. Study of the light plot in Figure 20-9 reveals PAR-64 fixtures as well as a 6 kW HMI (a metal halide lamp) for effects. This show definitely used theatre lighting, video lighting, concert lighting, and a lot of finesse.

What we really have with these corporate shows is not just a set of isolated examples, but another specialized field of lighting. This one transcends theatre, video, and rock and roll. It is the best example I know of how *all* theatrical lighting techniques can and should be used to produce the finished design. It clearly shows how we cannot close our eyes to cross-media applications and why it is so important to look at creative use of lighting in the broadest possible way.

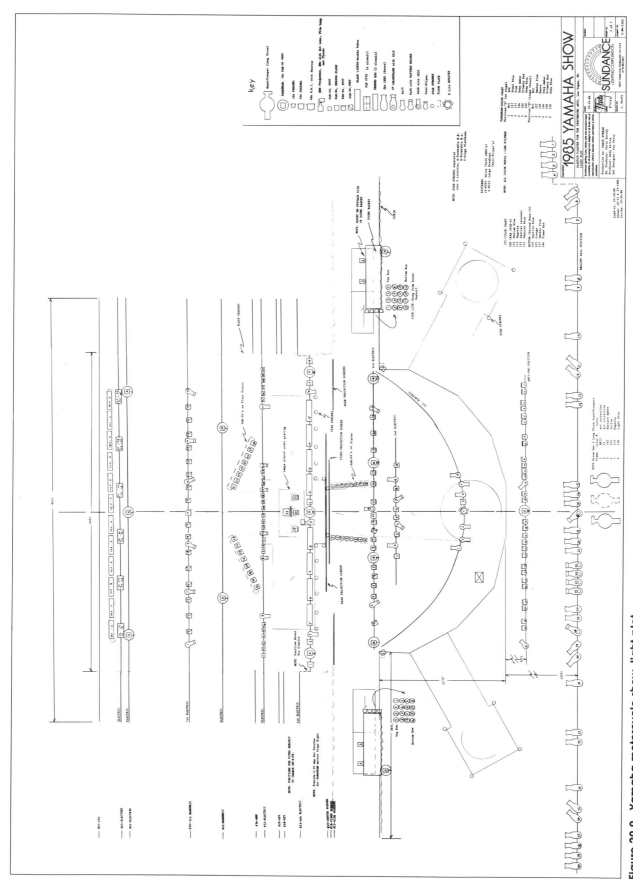

Figure 20-9 Yamaha motorcycle show, light plot

21

Open-Faced Fixtures for Theatre

Although this book has described concert lighting techniques and how they can be used in video, film, and theatre, perhaps an example of how film/video lighting can be used in theatre to solve a problem will help to show the full circle of the unconventional use of lighting that is developing in the lighting world today. We are no longer isolated and branded as "theatrical designers" or "concert designers" or "lighting directors." To work in today's lighting market, you must adapt and work in several, if not all, media. To use equipment to serve your needs rather than as you were taught is becoming more and more necessary. The following is a rather offbeat example of how the separate fields were joined together on a project.

Open-faced lighting fixtures designed for quartz lamps have been accepted for motion picture and television production for the past twenty years. The field that has not exploited this significant technical advance is theatre. Some older open-faced fixtures such as the beam projector and the scoop have been used, but they are a small percentage of the total complement of theatre fixtures normally used. The quartz lamp has become a standard even in theatre, not for its higher color temperature, but for lumen efficiency at a constant Kelvin temperature.

Actually, there had not been a technical breakthrough in fixture design since the ellipsoidal reflector spotlight was introduced in 1932. Reflector designs for theatrical fixtures have not changed appreciably for years. Certainly there have been advances in effects projectors and modifications in housing designs, such as the Parellipsphere by Electro Controls Inc., Mini-ellips, and the various computer-controlled moving light fixtures. New lamps such as the HMI (Metal Halide Iodide), CSI (Compact Source Iodide), and HTI (Metal Halide Arc) are new sources but for the most part they are still packaged in the standard Fresnel or followspot housings. Because of the high Kelvin temperature and selective spectrum range of these lamps they are of limited use in our current group of fixtures. However, they should not be overlooked and several very inventive uses have been displayed in productions recently. Currently they see the most service in theatre as followspot light sources replacing the carbon arcs. The HTI appears to be emerging as the lamp of choice for moving lights and smaller followspots.

In the overall picture, the open-faced fixture has had impact on several medias but not on theatre. The open-faced fixtures designed originally by Berkey-Colortran Inc. earned them an Academy Award ". . . for significant technical contribution to the advancement of motion picture lighting." Almost thirty years have passed since they introduced the first fixtures using the quartz lamp in a lensless fixture housing. As far as I know, no one has undertaken any extensive experimentation into their possible use in theatre.

A Theatre Experiment

The basis of my experiment, which is the subject of this chapter, was to determine:

1. lighting design possibilities in theatre
2. reaction of actors to the open-faced fixtures
3. a developmental basis for future design criteria

The technical styles of television and film production do not often require a clean hard edge, with minimum spill light. For this reason, a fixture has not been designed to meet this requirement of both hard-edged lighting and minimal spill light. If a hard edge is needed, the ellipsoidal reflector spotlight is still used.

For my experiment in theatre lighting with open-faced fixtures, I thought I should first pick a play and a theatre that would lend themselves to a production style that requires a minimum of precise lighting control. Second, a nonproscenium theatre was required so that we would not have to deal with the light spill on the proscenium.

The play I found was a comedy set on a three-quarter thrust stage. The production had opened a few weeks earlier, but the producer was not happy with the lighting. The offer was made to relight the show with equipment loaned by Joe Tawil, president of Colortran Inc., if he would agree to our experimentation. The scenic design employed one set that was altered between acts to represent two apartments, one above the other, in a New York tenement building. The biggest problem was to take down the previous lighting and replace it with new equipment while the production was still running.

Lighting Design

The production had originally been illuminated with seventy-four 150 watt R-36 lamps using hoods, snoots, and two-leaf barndoors. The complement of lighting equipment was completed with a boxspot (250 watt) and four 3 in Fresnels (150 watts each). A total of 11,950 watts illuminated the stage.

Readings of the footcandle levels as they were found and then re-checked after the open-faced fixtures were installed are shown in Figure 21-1. The arrows indicate the direction of the light when the reading was taken. It is obvious that the R lamps had contributed to a spotty and uneven job. The R lamps (built-in reflector in lamp) do not offer any control or flexibility and to use them as a sole source was not a good idea even in this open staging. Also note that the audience seating was measured to determine if the spill often attributed to the open-faced fixtures would be extensive. As you can see by the illustration, it was not.

After the usual meetings with the producer and director, the old lighting was struck and the open-faced fixtures hung according to the plot shown (see Figure 21-2). Because of the size of the stage, 25 ft × 12 ft, and the lighting angles we had to deal with, I decided to have a relatively flat wash of light from all three audience sides. With the height of the grid at ten feet, and the audience sitting under part of it, only very shallow angles were possible, thus seemingly adding to the spill light problem (see Figure 21-3).

The concept was to evenly wash the acting area and then accent areas with a brighter light source. This would create a highlight on

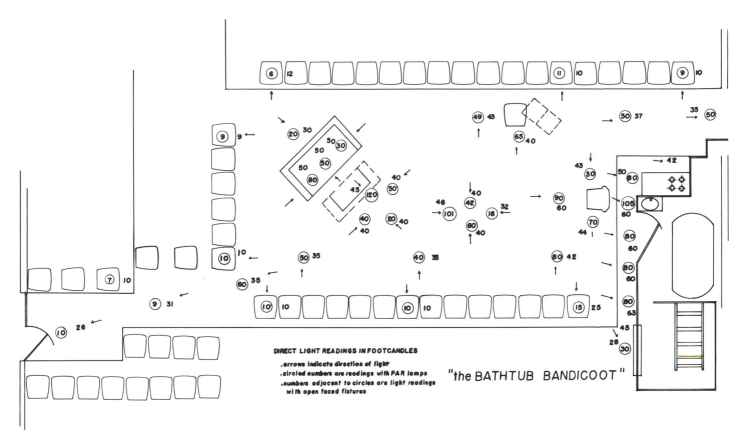

Figure 21-1   Footcandle readings before and after relight

DIRECT LIGHT READINGS IN FOOTCANDLES
.arrows indicate direction of light
.circled numbers are readings with PAR lamps
.numbers adjacent to circles are light readings
  with open faced fixtures

"the BATHTUB BANDICOOT"

the actors regardless of the viewing angle. The fixtures were kept to a minimum while a dimensional effect was maintained.

The position of the audience in relation to the hanging position of the lights added another problem. The Multi-Broads turned out to be focused in such a way that their reflectors were distracting to someone sitting in the front row. This was created by the physical limitations of the theatre, but I had to solve it. Because there was no lens, the reflectors seemed to sparkle. Spun glass, a diffusion media foreign to theatre but not to film and television, was placed in front of the fixtures. It diffuses the light rays so much that the problem of adding light to the audience had to be considered. As it turned out, I kept the light on the audience to under 10 footcandles of ambient illumination (see Figure 21-4). The only exceptions to this were four seats which recorded 25 footcandles. However, this was not due to the spun glass, but rather to the position of actors at the window area.

Fixtures Used

By using the Multi-Broad, which is a focusing broad-fill lighting fixture somewhat similar to a scoop, I achieved the flat overall light level of 40 footcandles. The Multi-Broads were double-hung for a color change to indicate the shift of action from one apartment to the other required in the script.

The Tru-Broad was used for back light and for one area coverage as indicated in the light plot. These units also function as a broad-fill light source but are more compact than the Multi-Broad.

The acting areas were then keyed with a light source that was placed to the downstage right side of each area. For this we used the Vari-

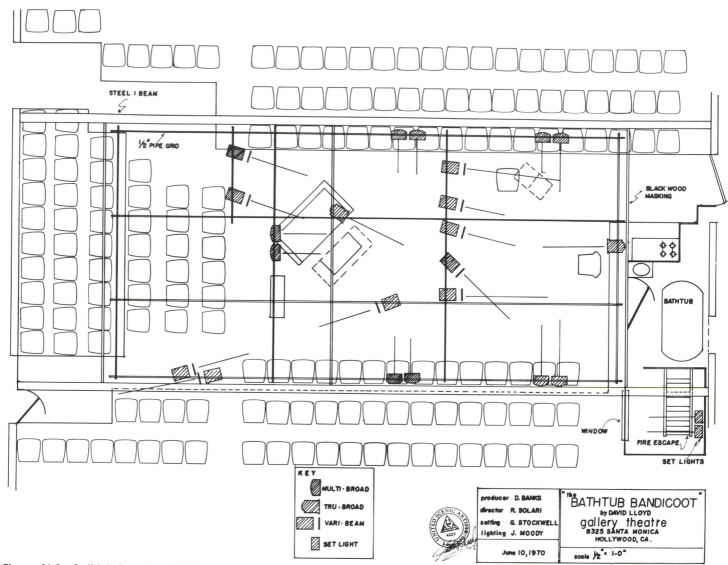

STEEL I BEAM

½" PIPE GRID

BLACK WOOD MASKING

BATHTUB

WINDOW

FIRE ESCAPE

SET LIGHTS

producer D. BANKS
director R. SOLARI
setting G. STOCKWELL
lighting J. MOODY

June 10, 1970

"the BATHTUB BANDICOOT"
by DAVID LLOYD
gallery theatre
8325 SANTA MONICA
HOLLYWOOD, CA.

scale ½" = 1-0"

Figure 21-2 *Bathtub Bandicoot* light plot with open-faced fixtures

Figure 21-3 Grid with open-faced lighting
(Photo by James Moody.)

Figure 21-4   Production photo showing proximity of audience
(Photo by James Moody.)

10, which is an open-faced fixture used much like a Fresnel. This fixture is also focusable and has rotating barndoors. Two Set Lights were placed behind the window for separate night and day effects. These fixtures, designed specifically for television, are for washing a drapery or backing wall with an even sheet of light. Their compact design allows for it to be placed very close to a wall and still produce a flat field.

The only lighting problem that would have required an ellipsoidal spotlight was at the aisle exit. Only the downstage left exit was used by actors and the door at the end of the aisle was used as part of the acting area. By adding extensions onto the standard barndoors on an open-faced fixture I was able to get a clean enough edge to work for this area.

The completed light pilot was composed of the following equipment:

Quantity	Fixture	Use
2	Set Light	Effects, night and day in window
5	Multi-Broad	Broad fill full stage coverage in cool color
5	Multi-Broad	Broad fill full stage coverage in warm color
2	Tru-Broad	Back light, center area key light
10	Vari-10	Area key light and aisle light

All fixtures used 500 watt lamps. A total of 9000 watts of light illuminated the stage at any one time. This was considerably less than the 11,950 watts needed for the R lamp fixtures. The light levels and coverage on the stage before and after relighting gives an idea of the difference produced both in light level, coverage, and efficiency of the equipment used.

Heat was also a consideration. Because of the wattage of the fixtures, several of the production personnel who were unfamiliar with the equipment expressed the fear that it would be too hot. Temperature readings before and after the installation of the open-faced quartz fixtures showed no heat rise.

Cast Reaction

The reaction of the cast to this new lighting was of major interest. To record their reactions the following questionnaire was devised:

1. Were you well covered in all of your positions on the stage?
2. Did you feel any abnormal discomfort from the lights?
3. Did you get spot blindness from looking into the lights?
4. Were the fixtures themselves distracting?
5. What was your first impression of the difference between the new lighting and the old?

Guy Stockwell was the most articulate of the cast and gave many impressions and comments during the run of the show. His feelings summarize those of the rest of the cast:

1. It was a well-illuminated stage.
2. There was no discomfort.
3. He had no problem looking out into the audience past the lights.
4. The fixtures were not distracting to him.
5. His first impression was "sparkle."

This last remark was expressed by several audience members as well: there was a snappy, sparkling quality to the lighting. Obviously this is a quality I would want for a comedy no matter what fixtures had been used, but it showed that it could be achieved with these fixtures. The color temperature of the lamps did improve the color rendering of costumes and sets.

A part of the credit should also go to the key light effect on the area lighting. Here is another example of cross-media, unconventional use of lighting. This was what made the actors stand out so well and added the needed depth and separation on this small stage.

The spun glass placed in the Multi-Broads worked as planned and no negative reaction was received from people in the first two rows who might have been bothered by these lights. With a little more height, no spun glass would have been needed at all, thereby increasing the overall light level another 40 footcandles. The additional light was not required in this case anyway, but on larger stages the increase in distance to the subjects would require the higher initial lumen level.

Summary

The experiment was a success for two reasons. First, we proved that the open-faced fixtures can meet some of the physical requirements of theatre, and second, they present no problems for actors.

This was an experiment and was hardly the ultimate in "theatre" lighting design, but the open-faced fixtures did what I expected of them. More development will be required to produce totally new designs based on the principles I learned from this experiment. There is no question that an advance in technology is long overdue in theatre lighting fixtures design. There are many techniques, materials, and concepts being generated in related and unrelated fields. It is a shame that advantage is not taken of these new tools.

Economics may retard the still fledgling attempts to improve lighting. Efficient, lighter, more compact, and better designed lighting fixtures will cost more. The question is whether there is a desire on the part of the theatre community to change, no matter what the cost. I hope that research is not slighted because of fear of being the first to pioneer new ideas. Certainly the creation of the computer-controlled moving lights by new nontraditional companies should be a beacon for others to develop new approaches to theatre lighting.

V

Afterword

22

Concert Designers: Art versus Business

Up to this point, you have been given only one designer's thoughts and opinions. Now I want to share with you the feelings of several concert designers who have influenced the media with their unique personal style and imagination: Leo Bonomy, Chip Largman, Jeff Ravitz, and Chip Mounk.

Leo Bonomy, Carnegie-Mellon University graduate, started at the bottom and worked his way up to tour designer. He ultimately designed for his old bosses, Chicago, and has been responsible for designs for Elton John and The Beach Boys. He is currently on the staff at the Universal Amphitheatre in Los Angeles.

Chip Largman got his basic theatrical training on the East Coast before moving to California. He tended toward designs for soft rock artists such as Olivia Newton-John, Neil Sedaka, Melissa Manchester, and Bette Midler. He is currently a production manager in San Francisco.

Jeff Ravitz has been a free-lance professional lighting designer for over thirteen years. A graduate of the theatre program at Northwestern University, Jeff has designed for some of the most highly visible rock stars in the world, including Styx, Air Supply, John Cougar Mellencamp, and Bruce Springsteen. He won the Performance Magazine Reader's Poll for Bruce's design in 1987.

Chip Mounk's career started when, as a high school dropout, he hit the road with Harry Belafonte in the sixties. Since then, he has been responsible for such well-known projects as Woodstock, the early Rolling Stones tours, The Rocky Horror Show, the Concert for Bangladesh, and shows for Rick Springfield. He is now with United Production Services, a staging company with national status.

I put the same ten questions to each of them. I wanted to discover how they felt about their relationship with concert lighting. They all approached the job with professionalism and concern for the client or artist they served.

Answers from Four Designers

Because these men are so busy, there was no time for formal interviews, and I have paraphrased their answers. I have added my own thoughts at the end of the chapter.

Q   What is the most important consideration when you first approach a design situation?

A   **Bonomy:** The size of the production and how much time you will have to mount it on the road—four hours, eight hours, or a separate day.

Largman: Personal relationship on an artistic level with the artist. I must feel that there is an awareness on the artist's part that I, too, am contributing artistically to the music.

Ravitz: Content and style. The type of music and performing style of the artist is the key to what he or she is trying to convey to the audience. This is a fundamental starting point to spark images for me.

Mounk: Space use. That is, what space the artist needs to perform. How many band members. Whether they can work in a configuration different from what they have used in the past.

Q What one area of the finished design will you not compromise?

A **Bonomy:** I am willing to bend, but with reservations. Trying to keep the concept, even if an individual aspect must be altered or dropped, is my idea of compromise.

Largman: The change of physical design, fixtures, structures, et cetera is possible, but I draw the line at show time. The actual cues are most important to me.

Ravitz: I will consider any suggestions, but I insist on control of the flow of the cues and overall balance of the stage look. I am the representative "eye" that has been hired to unify the look of the show, and I can tell if something works visually and is still true to the show's concept. Shows with too many chefs look like it. But I always listen and consider all input.

Mounk: Dropping colors. Once I have decided I need five colors to achieve the color mix I want, I do not feel I can drop a color and still achieve the effect.

Q What is the most important factor in a successful design?

A **Bonomy:** The onstage relationship of mood and color. Lighting the musicians to achieve an artistic representation of the music.

Largman: Focal consistency—all cues must tie together to create a flow. Light cues shouldn't detract from the artist.

Ravitz: A sense of dramatic orchestration that gives the show a visual beginning, middle, and end.

Mounk: Angle is most important to me. Achieving interesting angles of lights adds another whole dimension to the design.

Q How heavily does budget figure into your lighting design?

A **Bonomy:** Not too much. I work out how much can physically be put into the production, then set the budget. All the budget in the world will not get a production ready in four hours if it should take twelve. There must be a happy medium.

Largman: If there is a budget, it is good to know it; but it cannot overpower the initial concept. I would rather work toward creating a concept and then get management to put up whatever funds are needed.

Ravitz: It is satisfying to pull off a great design on a low budget. Creativity and resourcefulness abound. However, the minimum required is a realistic budget proportional to the size venues being played and the length of the performance.

Mounk: Design is all that matters. If you hire me, one of my conditions is usually an open pocketbook to complete my concept.

Q What tells you a design has succeeded besides your own satisfaction?

A **Bonomy:** When the audience is with the band. When lights, music, and set all come together for the audience.

Largman: When the design works from show to show. When the audience is not aware of light changes, but they react positively to the artist.

Ravitz: Audience reaction to the show. A bad design might actually stand in the way of total appreciation of the performance.

Mounk: When there is no mention of lighting in the press. The old theatrical adage, "When they go out humming the sets and lights the play is a failure" is true here to a great extent.

Q Who do you prefer to deal with: artist, road manager, or personal manager?

A **Bonomy:** The artist in the concept stage. After that, whoever is in charge of the purse strings.

Largman: Manager or production manager if there is one. But after the designer and artist have an understanding of the concept.

Ravitz: I prefer to deal directly with the artist on creative, conceptual matters if the artist is so inclined. Otherwise, the personal manager often takes a directional role. Ultimately, a meeting with all three is the best if a unified direction can be arrived at. Then the creative and the logistical can be considered at the same time.

Mounk: Accountant on the road, but only after the design concept is locked in with the artist's approval.

Q Will you adapt a design to changing venues or start all over?

A **Bonomy:** I like to use the unique features of halls to enhance the road design. The initial design is based on the type of halls the show is to play, but I adapt when necessary.

Largman: I will adapt the initial design up to execution of the first show. Then it is my responsibility to make changes only as I see fit. If the production changes halls, I will adapt to the new situation.

Ravitz: On a sizable tour, I will adapt an existing, successful design to a change of venue type, if possible. Even drastic changes would retain key elements.

Mounk: I work more in concept and space and leave the everyday road changes for differences in venues up to the road electricians. As long as they continue my concepts, the design can be adapted as necessary.

Q How important is formal theatrical education?

A **Bonomy:** There's a lot to be said for college, but discipline and experience are the true teachers. Lighting music is different from lighting plays, so it is not the most important factor in the background of a designer.

Largman: It gives a sense of the classic and a broader background to the designer. He is better able to deal with concept instead of just putting light on the stage.

Ravitz: A formal education has its place in learning certain basics in the most straightforward way. To learn where the craft has come from helps to guide where it will go. But, there are good and bad learning environments and good and bad students. One would have to get all one could from school—a sense of history, art, methods, organization, and discipline—and then go out and *do it*.

Mounk: I find college-trained designers are afraid to use hard, bold colors. They are usually not able to make commitment to design concepts. It is a lack of personal confidence.

Q Should a person go on the road as a technician before becoming a designer?

A **Bonomy:** Yes, to learn the very specialized touring structures and methods of packaging not taught in college.

Largman: Not critical, but it is important to have exposure to structures and fixtures.

Ravitz: Certainly, road experience opens a young designer's eyes very wide. There is no way to gain that sixth sense of what works and what doesn't without the road, both from first-hand experience and from comparing that to what every other show is doing.

Mounk: Definitely, unless you are also a structural engineer and logistics specialist.

Q What do you, as a designer, feel is the most important personal quality needed to be a successful concert designer?

A **Bonomy:** Confidence in yourself and ability to stand up for your concepts.

Largman: Know your limit and know where to get help.

Ravitz: A designer must be part diplomat, part scientist, and part artist. The designer is the leader of an entire department of the tour, so leadership and sometimes parental skills are necessary. One must also possess patience and confidence. But, in the end, it is a balance between being open-minded and standing your ground.

Mounk: Selling yourself. Once people have confidence in you, getting what your design requires is easy.

My View

Each of these men entered concert designing via a different path and achieved a place as a top designer. Several have already gone on to other media or executive positions and have been successful in these positions also. I believe that shows that they had more than just a design skill; it speaks very well for their managerial and business skills. My choice of these four men in no way was meant to imply that women do not do well as concert designers. There are very well known women in touring, including Marilyn Lowey in Los Angeles and Abbey Rosen in London.

In the limited time we had to speak, we were not able to get into a lengthy discussion. Therefore, I will fill in some points that were not covered.

When you first approach a design project, your most important consideration can be as varied as your own background. Just as color or space may be important, so can acceptance by the artist, band, and road crew of the fact that you are in command. I use command in the sense of having thoroughly prepared yourself to undertake the project. Do not let the crew get the impression that you cannot deal with problems or are not flexible and willing to work with them.

To compromise a design can be as simple as making the cyc green instead of blue-green because the artist likes green. We all compromise, daily, so why not in lighting designs. This is a medium where factors concerning sound, musicians, space, time, and money all come together. There are bound to be compromises with all of the variables in the game. Simply face them and do not feel that any problem or obstacle cannot be overcome. Sometimes you have to stand your ground, but in the end, it is the artist's show and whatever is done must satisfy him/her/them, just as in theatre or film where the director has the ultimate control. The truly final word is your paycheck: do good and you get one, do bad and they are few and far between.

A very important factor in a successful design is simply *you:* your own instinct and understanding of the problems. Your feeling for the needs and limitations of the artist is what makes each design unique. We work within a very tight time frame. Given enough time and money, the place to put lights, a stage, and crew, anything can be accomplished.

To me, budget is the thing. Sure, I like it when a client does not appear to care how much it costs, but I care. I know I will not be back if he does not make money. I prefer to deal with a client knowing that I have a ballpark figure to work from up front. I may exceed it, but at least I had a guideline. Maybe I am more comfortable in this area, because I owned equipment for a number of years and perhaps have a better sense of what things are going to cost.

A design is a success for me when the house stagehands tell me it was good. They see show after show and for them to take the trouble to come and find me after the show and compliment me means a great deal. Besides this, I also agree with the maxim Chip Mounk quoted concerning successful designs, so I hope people will not go out "humming the lights."

Naturally, the artist can lay down guidelines, but some do not. I have worked with clients where I have not met the artist until dress rehearsal. Then I take the advice of a good road manager who has been with the artist for a year or more, because he sees the show the artist never sees. He hears the manager's complaints about money and the artist's gripes. He is on the firing line. The show must be good but still make money or he is out of a job. He is the balance point between the artistic and monetary sides of the show. However, I do take his artistic comments with a grain of salt, because I have been misled by them before.

If I find out that a show I designed for arenas is going into Las Vegas, I want to start over. I may still use the ideas that worked in the original design, but it gives me a chance to take a fresh look at the concept.

I like young designers who come to me with a college degree under their belts, but professional experience will outweigh education. No school is currently offering advanced studies in this media. After school, put in time on the road. An old Air Force technical school sergeant once told me, "Okay, you know the book, but you learn the really important stuff like who goes for coffee when you get out in the field."

There is another aspect of working on the road as a technician. In several instances, I have had excellent theatre students work for me who simply could not adjust to living out of a suitcase. Doing a show for two weeks in one town is quite different from forty days of one-nighters. Always working against the clock to get the show up on time or get to the next town on time is physically and emotionally draining. Some people like to be free, with no base, others need a home. Ties break easily in touring.

The ability to not be afraid of making a decision is the most important quality a concert designer can have. This comes from being confident in yourself. Decisions are asked of us every day of our lives, but seldom with more urgency than at three o'clock in the afternoon in Des Moines with eight channels of the dimming blown up. Sure, the show will go on anyway but you have to make the best of it. So do something, anything, but take a stand and follow through.

23

Overseas Touring

When American artists decide to go on tour overseas, one of three approaches to their technical needs is employed. Most commonly, the lighting designer is told by the artist's management that all arrangements have been made through the local promoter and that a local supplier will provide equipment *as available*. You are expected to "make it work, and besides, this is not the United States!" In a second approach, you may be asked to submit a lighting plot, but from there on you are told that you must accept what they give you. Or an uncommon thing can happen. In the third scenario, you can actually get just about anything you request!

Touring Europe

If you are touring in Europe, you will find virtually everything that you need and are familiar with in the United States. Added to all the American-built equipment that has been exported to Europe are the English and European fixtures, dimming, and consoles. You have a candy store for designers.

Besides the English companies, several of the U.S. tour lighting and sound companies have branches in England; Lighting and Sound Design, See Factor Inc., Showco, and TASCO have affiliated firms operating in England and on the Continent at the time of this writing. Check with them or any other U.S. company early in your preparation for your overseas assignment.

These companies combine European and U.S. technology to create their systems. Therefore it is relatively easy to tour in England and Europe as far as the equipment is concerned. Just about any piece of equipment is available. However, quantities are limited and early commitments are usually necessary if you want to be assured of getting everything you want.

Cultural Differences

The cultural differences are much more of an adjustment for the designer. The customs, such as work habits, as they apply to the way Europeans take their breaks, meals, and even crew availability can vary widely throughout the Continent. That is not to imply that they are not as good as, if not better than, American crews. It is just that you must keep in mind that they were raised with a different work ethic and it can take some getting used to on your first overseas tour.

Language also comes into play. You can carry one of the handy translation books that cover theatre or film terms, but that will not be the total solution to the problem. Many promoters do hire English-speaking crews on the Continent; in some cases they are American servicemen or their dependents.

I find the best solution is to make sure the head electrician you get from your British equipment supplier is able to converse in French and German. That is not too much to ask as many Britishers do speak some of both languages.

Working through an interpreter is difficult because most are not familiar with the theatrical terms we need to have communicated. I also find that the interpreter ends up spending more time with the management personnel than with the crew. I have gotten pretty good at the few words and numbers you need to communicate and I usually find that enough people speak some English so that you can get your point across.

However, I did do a show in Berlin where it turned out only one crew member spoke a little English. We were playing the then new International Congress Centrum. The plan was to use only the house lighting crew and equipment for this corporate show. Because we had a two-day load-in scheduled, I agreed to try to work with the crew. We had a translator, but I knew she would be busy with other people most of the time (and I was right). I was told that the crew all spoke some English. The fact was that the facility's Technical Director was an American, but he could not be around all the time to lend a hand as he had a very large facility to run with several other theatres and convention spaces. So, I was left to my own devices.

With a lot of gesturing and the few words I knew, we got the show mounted. The real problem was in calling the followspots. What I contrived was to write out all the cues and spend plenty of time with my one English-speaking man going over each cue until I thought he understood. Then I would point to the cue that was next up and he would read the cue in German to the followspot operators and I would give the GO or Black Out (Hauptregler) or Fade Out (Ausblender) in my best German.

The crew tried very hard and did an admirable job. Of course we had two days of rehearsal besides the load-in time, so I could take this chance. On one-nighters, I would not recommend trying this at all.

Another case, also in Germany, came about in trying to convince the local officials that it was safe to rig a lighting truss from a catwalk in their beautiful and very old municipal theatre. Every official we talked to passed the buck up higher. Finally we had to call in the town's Burgermeister (mayor). The German promoter was acting as translator and go-between and they were all talking so fast that I could not understand what was going on except that it did not look like I was going to get my truss hung. I finally got them to stop long enough for me to motion that we should all go up on the catwalk so I could show them what we needed to do. Once on the catwalk I asked the promoter to tell them that all of us standing at the very point we wished to attach the truss weighed more than the lighting truss! We did the show as planned.

Responsibilities

In most European countries, especially in Germany, the technical director is personally responsible for the safety of the theatre. He can be legally held liable, fined, or jailed if there is an accident. Of very great concern to the authorities throughout Europe and England is the rigging. Many of the buildings are very old and highly suspect as to their actual load-bearing ability. Do not be shocked when a rigging plan representing all the rigging and bridle points and their respective weight loading is demanded in advance of the show. In point of fact, the Greater London Council (the GLC, as it is commonly referred to by Londoners) will request such a plan and they will need to approve it several days in advance or there will be no show. (Figures 23-1 and 23-2 are examples of rigging plots.

This is a practice I wish was more prevalent in the United States. I believe I have worked with the finest riggers in the world and safety is always a top concern, but even they would be the first to admit that many buildings have not had a safety test inspection in many years and that structural stress can change from day to day. Any touring designer or rigger who does not worry about each building he or she enters is a fool. And that high level of concern is what keeps us safe.

Power Supply

Power problems are not serious if you are using one of the British concert equipment suppliers. They are prepared for the different power setups in each of the countries and have adapters to do the job. However, if you insist on bringing your own lighting console from the United States, make sure you request a transformer to plug it into or you will be out of luck. Many of the rock and roll lighting consoles do have a built-in switch, especially those made in Europe, so that they will work on 110 and 220 volts. Any band gear with motors or fans will need 50 cycle adapters, which are expensive, so it is better to rent that equipment (electric organs, computer-based keyboards, etc.) in England.

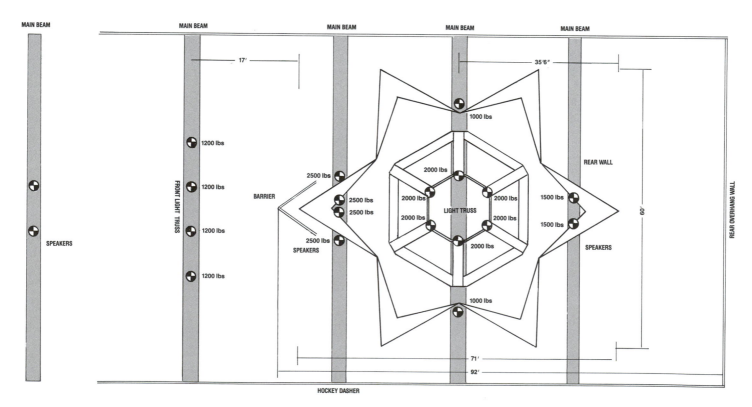

ROLLING STONES TOUR 1975
STAGE PLAN

BUFFALO MEMORIAL AUDITORIUM

Figure 23-1 Rigging plot, Rolling Stones tour
(Plan by Brannam Enterprises.)

Figure 23-2 Rigging plot, Bette Midler tour (Plan by Brannam Enterprises.)

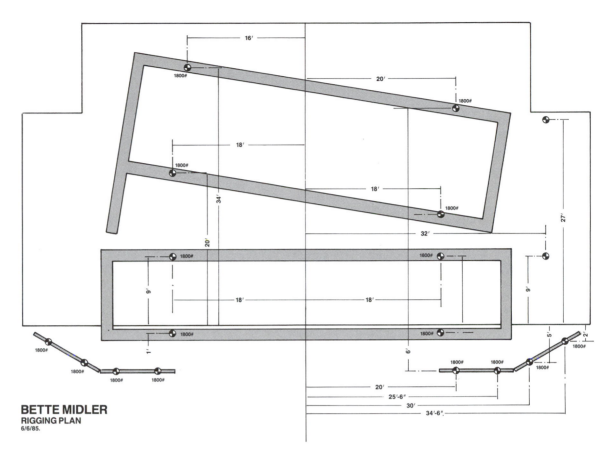

BETTE MIDLER
RIGGING PLAN
6/6/85.

Touring Japan

The Asians are late to enter the concert market as packagers of tour systems. Led by the Japanese, Asia has become a big market for American and European artists. Because of the extreme cultural and language differences, which are a seemingly insurmountable wall to the Occidental, we will concentrate on this market because it presents the biggest difficulties.

In the past few years a burgeoning group of companies has appeared in Tokyo, eager and willing to serve the foreign lighting designer's needs.

The Lighting Industry in Japan

The cultural differences separating our two countries would take volumes to explain, even superficially. This is acknowledged even by Ken Lammers, who was born in Japan of American parents, and has spent all but three years of his life there and worked his way up to become a manager in one of the largest Japanese lighting companies. Recently he started his own company to supply bilingual backstage coordinators. He would be the first to say that he will never be completely "Japanese."

Inadequately stated, the Japanese view theatre as an art refined to its most simple common denominator. Light is there to make it possible to see and is not a dramatic part of traditional Japanese theatre. Therefore the light should be bright, even, and clean so the audience can see the subtle lifting of an eyebrow, or the twitch of a finger during a scene from a Kabuki or Noh play. Form and economy of movement are the keys in these ancient dramatic arts. The schools do not teach

Western theatrical lighting techniques and the theatres are not equipped for "Western" lighting designs.

Theatres and Concert Production in Japan

All of the major theatres, some 2,000, were built after World War II. Most are government owned, whether by the city, school, prefecture, or federal authorities. These were built for community use and each was designed as an all-purpose theatre equipped to handle all the performing arts. Most theatre systems were standardized for time efficiency. Because the theatre has to serve the whole community equally, no one person or group can book the hall for a long period. This means that most shows must start to set up on the morning of the show. Other than a few privately owned theatres, it is hard to get more than three or four hours to set up and focus before rehearsal. This is why the zoom type ellipsoidal, the plano-convex (box) spot, and Fresnels were used and are still there today.

The use of color media and more complex control entered Japan in two ways: Western-style theatre and American modern dance. Taking their cue from the likes of Martha Graham and others, the modern dance of Japan is very experimental and innovative, and this sparked new uses for lighting.

In the early 1960s there were no equipment rental houses except those that catered to the movie industry. When television came out of the studio and began shooting in the theatres, a few lighting rental companies started to appear. This was when the six-channel, 6 kW dimmer packs were first introduced to the market.

When rock and roll concerts came to Japan it was a surprise to the Japanese that people wanted to take a full lighting system into a theatre that already had one. For a long time artists had to use the house system unless they brought the whole thing with them from overseas. The Budokan in Tokyo opened its doors to concerts and was the biggest influence on touring equipment. Trusses started to be used, although they were constructed of steel. Most of the rental houses were using the American-made 120 volt lamps because they were economical. The lumen output and color temperature were bad because they were being used at 100 volts. In the late 1970s, one of the lighting rental companies, Kyoritsu, made the first 100 volt PAR-64. Even with this new lamp, most shows did not carry their own dimming and console. This was because most of the halls did not, and still do not, let you into their main power. This means that even if the hall has enough power for the show you are forced to rent a generator. Most of the promoters have accepted the need for a generator as part of the production cost, so most tours from overseas can have their own control consoles and dimming now. A few Japanese artists tour with portable systems, but for most, it is still common to use the theatre equipment.

The 1980s have seen a big change in concert production. In the following pages I will detail some of them.

Lighting Companies

In Tokyo alone there are now about 110 lighting companies plus some theatrical groups that have lighting departments registered with the Lighting Engineers Association of Japan. But most of these companies only have crews and not equipment. Of the companies that do, only six or seven do concert touring for Western performers. Listing them in alphabetical order: Kawamoto Stage Lighting Ltd., Kyoritsu Ltd.,

Lighting Big-1 Inc., Lighting Version Ltd., Sangensyoku Ltd., Sogo Butai Inc., and Tokyo Lighting Inc.

There are two or three rental houses that cater to the remaining companies, which do not own their own equipment. There are no full-service companies that provide sound, trucking, rigging, staging, and lighting, although Kyoritsu Ltd. comes closest with the addition of their subsidiary companies that provide sound, stage carpentry, and video services.

Lighting companies do not take care of rigging of trusses, drapery, or risers either. This means that there will be more than one company needed to supply the tour.

Power

Japan is the only country in the world that uses 100 volt AC power. It is also useful to know that northern Japan is on 50 Hz (Hertz) while southern Japan is on 60 Hz. Also there is no law that requires each light to be grounded. Grounds are required for installations up to the main breaker box, but the individual outlets do not need to be grounded. If you bring equipment that needs to be grounded make sure you do it yourself and do not depend on the house having a ground. Check before you connect anything.

Part of the reason for not simply putting a company disconnect in the theatre has to do with the electrical laws of Japan. Any temporary hook-up that is over 100 kW (about 300 amps) needs to have a qualified electrical supervisor on hand. It is easy to see that this would become expensive just for the six or seven shows a month that would need the service. This also applies to generators, but as long as there is an operator from the generator company on hand, they seem to allow it.

Generators

Due to stringent noise pollution laws, generators must be extremely well sound insulated (blimped). The most common size is around 120 kVA (kilo-volt-amperes). The largest available are around 300 kVA. Taps can be changed on all sizes so the voltage between the hot legs is either 440 or 220 volts and then varied as much as plus or minus 50 volts. In this way all needed voltages can be obtained. Also the cycles can be varied between 50 Hz and 60 Hz.

Equipment Availability

PAR-64 Lamps Even after the 100 volt PAR-64 was introduced, many companies continued to use the 120 volt lamps because there were only narrow beam and medium floor lamps available at the lower voltage. If you needed the very tight focus of a very narrow lamp and wanted it to be brighter than the 100 volt medium flood you had a problem. In 1984 a 500 watt very narrow lamp was produced. This light is so efficient that it is brighter than the 120 volt 1000 watt lamp when used on their 100 volt systems. Now there is the whole family of very narrow, narrow, and medium floods.

Ellipsoidal Fixtures The rash of Broadway musicals coming to Japan the past few years has caused the Japanese companies to purchase a great number of ellipsoidal fixtures. The most common are 6 × 9 and 6 × 12 with 100 volt lamps. There is also a Japanese ellipsoidal with a zoom lens. Its beam spread is just about equal to a 6 × 16 at the narrow end of the focus. If you need 6 × 22's it is best to bring them with you. Although the inventory has grown, only a few companies have them. Subrental is common, but leave enough lead time to allow for negotiating between companies.

Projection Equipment Even though the RDS projectors, known in America as the Scene Machine, are made in Japan, the availability of the effects heads is limited, especially for projection of painted glass slides. If you need to use a remote system and change speed or slides during the show it may be best to bring your own. You can get 35mm slides and film projection equipment, and because most of it is imported, the make of equipment you are used to in the United States is probably available. Video projection equipment is also available.

Other Lighting Equipment In general all sizes of Fresnels, striplights, and plano-convex spots are readily available. The Japanese-built equipment is excellent and compares very well with the U.S. and European equipment, but you need to work with it before you commit a full design to unfamiliar equipment.

Followspots Most of the theatres are equipped with 2 kW Xenon units. There are usually four in a theatre. For performances in other types of venues, the rental companies will supply your needs. It is interesting to note that most Japanese operators disdain the color changers and place colors manually. Mr. Lammers said that he gets quite a reaction when he tells the foreign designers that their show was just done this way. The smooth and accurate operation they achieve, even on fast rock shows, is undetectable to most people.

Trusses and Hoists Aluminum trusses are available but quantities are limited. Most are imported, but some are made in Japan as copies of European designs. Hoists are Japanese-made because of voltage differences from the U.S. and Europe.

Lasers There are no federal laws on laser use, but some local restrictions do apply. They are available, and the promoter can put you in contact with a specialist.

Trucking All trucks are eleven-ton, straightbed frames, 36 ft long. Few tractor-trailer rigs are available in Japan and no air-ride boxes are available. Luckily, all Japanese roads, even in the countryside, are excellently maintained.

Crew and House Staffs

There is no union of stagehands in Japan, although some of the privately owned theatres have in-house labor unions. The government-owned halls will have a two-person lighting staff to act as supervisors only. Between shows they maintain the facility in top form. The crew needed to load in and run your show will be contracted from the outside. This is usually done by the promoter and the theatre has no say in which firm is hired.

If the show is going to travel outside of Tokyo the key people will travel with the show and local crews will be picked up at each stop. This is arranged by the lighting companies through regional offices or agreements they have with local companies.

Crews get no set breaks or minimum guaranteed hours. They are paid a flat rate for the day with a bonus if the work goes through the night. The full crew will stay through the load-out, even if that is the next day. Most Japanese have the attitude, "Let's get the work over with and then rest." Most crews will want to work until the focus is finished before taking a break even if that means no lunch. Japanese promoters are not in the habit of feeding the Japanese crews. If they do, make sure special meals are ordered for the American and European crew members. The typical Japanese worker's meals will not be palatable to most outsiders, even if they think they have eaten true Japanese foods in other countries. The diet is very different and takes some getting used to before you can eat whatever is available.

Payments and the Promoter

The promoter normally contracts for the lighting, sound, trucking, rigging, and staging for a tour rather than the artist's manager as is done in the United States, South America, and Europe. There are no Japanese promoters who enjoy an exclusive in any facility. The lighting and sound companies will have ties with different promoters and there is not the bidding situation encountered in the United States. In the past it was hard to convince the promoter to take a lighting system on tour. Now almost all tours will travel with a full system.

Just as in the United States, the production costs are mounting largely due to bigger and bigger designs. It is common to see a promoter write into the contract that the artist is limited to a fixed cost for production and if they want more, they pay the promoter back for the additional cost. Lighting plans do get cut down during the production meeting between the Japanese promoter and the equipment company, so be prepared to convince the artist or their manager to pay up or be willing to cut back the design. The cost for equipment and crew runs about one third more than in the United States.

Theatres and Other Halls

Most theatres are under 3,000 seats, but they have very wide stages by U.S. standards. A 100 ft wide proscenium is not uncommon, but the working depth is very shallow. This is due to the traditional Kabuki theatre's needs. The larger halls are usually gyms, which hold 5,000 to 7,000. There are about four halls in all of Japan that hold 10,000. Recently artists such as Madonna and Michael Jackson have used stadiums because of the greater demand for tickets.

As mentioned before, most halls are government run and thus there is often no discernible logic to the rules they impose, even discounting the cultural differences. A bureaucrat is a bureaucrat the world over.

Booking a hall must be done well in advance, usually one year. This means that many times a promoter will book a hall even if he has no artist scheduled at the time. The halls, being government owned, have to give an equal opportunity to all who apply for dates, thus it is impossible to get more than a couple of days back to back. If an artist wants to or ticket sales demand longer runs, the promoters will often get together and shift schedules when possible, but it is not a given that an agreement can be made, so do not count on an extended run.

Each hall has a different starting time. But the common rule is a 7:00 or 7:30 P.M. show. Halls will want the show to end by 9:00 P.M. This is not only because of the work schedule of the hall staff, but most of all for the audience. Most people come to the theatre directly from work or school. Public transportation is used by the majority of people, who then face about an hour to an hour and a half ride one way, so they need to be on their way by 10 P.M. It is common to see people leave a show before the last song to catch their train, not because the show was bad.

One other reason for the early stopping time is noise. The legal penalties are very harsh on the promoter for violating the noise pollution laws. Because homes are built on all available land, the loading doors probably look directly into someone's bedroom. Noise after 11 P.M. is not tolerated and the theatre will get complaints immediately if this is disregarded.

Rigging

If the lighting system can be arranged on straight pipes it will make life much easier for everyone involved. Theatres do not like to see trusses hanging from their grid, even if they are lighter than the equipment they replaced. If you have to bring a truss be prepared to ground support it. Fire curtain laws are strict, so no box truss configuration can be hung unless it is behind the curtain line. There are no halls as of 1987 that allow hanging points in front of the proscenium. The only answer is to move the artist back away from the audience, which will not make the performer happy.

The gymnasium type halls are gradually allowing rigging but each has different weight restrictions and most seem to be ridiculously low compared to the U.S. limits.

The Budokan, one of the best known and biggest venues in Japan, has gone through a few stages regarding rigging. The hall was built for the 1964 Olympics and has a concrete false ceiling.

One foreign group did rig and there was some damage, so rigging was banned for some six years. After long consultation with the architects who designed the building, rigging points have now been installed. Each point has a dynamometer in it that gives an instant readout. If it goes over the weight restriction a warning bell will ring and work stops until it can be fixed. Each point has a different weight limitation depending on how the other points are being used. Only designated bridles and Japanese-made cable hoists are allowed. A computer controls all the points and there are ten patterns that it will allow. You must design the grid to fit one of these plans. The total weight allowed is 7.2 tons but because of the weight distribution restrictions, only about 5 tons can be over the stage area. If you ground support, there is no restriction.

Pyrotechnics

Pyrotechnic regulations are up to each hall, city, or prefecture fire marshal. Permission can take up to a month and advance notice may be required in writing. On the day of the show an inspector has the power to disapprove, even if you comply with the written order that was issued.

Business Ethics

The way business is conducted in Japan will be a puzzle to you. We have all heard about the Japanese being interested in "saving face" above all other considerations. The formality of business entertainment is also strange to the Westerner. In Japan, 4 percent of a firm's gross revenues are tax deductible for business entertainment.

It is more than being friendly to ask a client out for dinner and drinks; it is part of the normal business day. I have often seen a businessman, briefcase still in hand, leave a bar at 9 P.M., bid goodbye to a group of similarly briefcased men, and head for the subway train to go home for the first time that day. It is an expected ritual for corporate executives and it is a great offense to decline such an invitation.

Another facet is the pecking order within the firm. If your boss (usually translated as Superior) is to have dinner with an important client, you could be asked to attend. And if you supervise a department, your assistant will also attend. All this is a show of status and position, not because they need the advice of their employees.

This brings up another rule of business in Japan. Often a question is asked directly of a technician and a clear answer is not given. Their system requires that when a superior is present, the question must be answered by the higher ranking person. Westerners usually view this as subservient behavior, but you should realize that it all fits into the Japanese sense of order and respect for elders and those of higher position in business.

Respect for Other Cultures

Touring in Japan or any foreign country is exciting and challenging. Before you condemn methods and work habits you encounter overseas, look to the social and economic structure of the individual society. We in the United States have cultural links largely in Europe, so Asian travel can appear even more unusual. No matter what country you travel in, remember that they are on home ground and you should respect their beliefs and ways of conducting business. Take time to prepare for your overseas tour, not only technically, but emotionally and culturally as well.

24

Postscript

The field of concert lighting has moved well beyond the narrow label of rock and roll lighting. In the past few years we have seen methods originally developed for tour lighting adopted by or adapted to dance, theatre, Broadway, bus and truck companies, location television, and yes, even opera. Film is the latest to launch a guarded investigation of these techniques.

The use of the PAR-64 fixture and freer use of bold color media attests to the acceptance of some of these approaches.

Early concert-style rock and roll performances on television were accepted by the viewing public so well that the television people took notice even if they did not jump in with both feet. I introduced the use of color projected on people rather than limited to the cyclorama on the *Don Kirshner Rock Concert* series in 1973. Now the PAR-64 finds wide use on location video and even in network news lighting. It is becoming rare to find a television studio where a PAR-64 or two is not in use.

I am not advocating the ouster of the venerable Fresnel, but I am an advocate for continued change. Change happens in two ways: through new inventions, and, probably more realistic for theatre and all its sister media, through the borrowing of techniques and equipment. Lighting has never been a heavily financed area of research in the theatrical arts, and therefore we must take what we can from other sources.

Concert designers and their equipment suppliers have taken found space and created "theatre." When people are able to reach beyond the rock and roll label they can see what real advances have been made, both in design potential and in the high-tech electronic explosion spearheaded by the intelligent lights.

I do not like all rock and roll music, just as I do not like all opera, but it is time we investigated the *art* of this media, without being hampered by our own prejudices against the music itself. Take the good elements and discard the bad; improve techniques and adapt them to other areas. This is the essential character of theatre, the great adapter.

This book reveals only a small part of what concert tour lighting has to offer. The Bibliography gives the names of magazines that regularly print articles about concert lighting. For those who need to broaden their basic knowledge of lighting there is a list of books I feel are essential to making use of the full potential of the material in this book. I hope you will obtain these books and add them to your store of knowledge.

Bibliography

There are currently no books on the subjects covered in this publication, only articles appearing in technical and professional periodicals. The list below is given as supplemental reading on concert lighting and other lighting techniques. Reprints or back issues can be obtained directly from the publisher.

The books that are listed are given to help readers broaden their knowledge of the other design media discussed in this book. Crossover use of techniques is constantly being tried in other media, so a better understanding of all media is necessary to compete in the lighting world today.

Periodicals

On Stage
On-Stage Publishing
6464 Sunset Blvd.
Suite 570
Hollywood, CA 90028

Tabs (out of publication)
Rank Strand Electric
P.O. Box 70
Great West Road
Brentford
Middlesex TW8 9HR
England

Lighting Design and Application
The Illuminating Engineering
 Society of North America
345 East 47th Street
New York, New York 10017

Cue
Twynam Publishing Ltd.
Kitemore, Faringdon
Oxfordshire SN7 8HR
England

Theatre Crafts
Theatre Crafts Associates
135 Fifth Ave.
New York, New York
 10010-7193

Lighting Dimensions
Lighting Dimensions Associates
135 Fifth Ave.
New York, New York
 10010-7193

Light and Sound International
John Offord Publications Ltd.
12 The Avenue
Eastbourne
East Sussex BN21 3YA
England

Pro Light & Sound
Mountain Lion Publications
5302 Vineland Avenue
North Hollywood, CA 91601

Strandlight
Strand Lighting Ltd.
P.O. Box 51
Great West Road
Brentford
Middlesex TW8 9HR
England

Books

Electronic Cinematography
Mathias, Harry, and Richard
 Patterson
Wadsworth Publishing Company, Belmont, Cal.
1985

Theatre Words
Luterkort, Ingrid, ed.
Nordic Theatre Union, Solna,
 Sweden
1980

Bibliography

Lighting the Stage
Bellman, Willard F.
Harper & Row Publishers,
    New York
1967

*The Techniques of Lighting for
    Television and Motion Pictures*
Millerson, Gerald
Focal Press, Stoneham, Mass.
1982

The Lighting Art
Palmer, Richard H.
Prentice-Hall Inc.,
    Englewood Cliffs, N.J.
1985

TV Lighting Methods
Millerson, Gerald
Focal Press, Stoneham, Mass.
1982

Cinema Workshop
Wilson, Anton
A.S.C. Holding Co.,
    Hollywood, Cal.
1983

*Concert Sound & Lighting
    Systems*
Vasey, John
Focal Press, Stoneham, Mass.
1988

Stage Rigging Handbook
Glerum, Jay
Southern Illinois University
    Press, Carbondale, Ill.
1987

Glossary

Accent A design technique in concert lighting that concentrates the illumination at strategic points on the stage to punctuate the music with heavy color or intensity.

ACL (Aircraft landing light) A PAR-64 design with a very narrow beam pattern which operates at 12 volts and was originated for jet aircraft, then adapted for rock and roll.

Air light The use of light beams to create patterns and design elements. The patterns may or may not be focused on the artist.

Aircraft cable (wire rope) A cable made of stainless stranded steel wire of high tensile strength used for hanging trusses, sound, and scenic elements where house pipes are not available. Theatrical rigging usually uses ⅜ in or ½ in diameter cable.

Back light Illumination of a subject from the rear to produce a highlight along its edge.

Box truss Structure built of aluminum, steel, or chrome-moly, designed as a portable support of lighting fixtures but sometimes used for support of drapery, scenery, or sound equipment.

Bridle In rigging, a system for utilizing more than one hanging point to distribute the load.

Chain hoist Electrically operated lifting device usually using chain; often referred to as *chain climber*.

Chase To sequence channels of dimming in a timed, programmed manner.

Color temperature The color quality of light, measured in degrees Kelvin, relating its spectral distribution to that of a standard ''blackbody'' radiator.

Contract rider A supplementary agreement between the promoter and artist, having its basis in the original contract and incorporated by reference to the original contract. Normally it contains specific physical requirements of the artist's current show such as stage size, power, dressing room needs, food, etc.

Contrast range The brightness ratio between the lightest and darkest tones in a scene.

Cyclorama A vertical surface used to form a background. Generally it is monochromatic and is of either hard wood finish with a fabric covering or it is a heavy cloth, often muslin, stretched vertically and horizontally to create a smooth surface.

Depth of field The zone or range that still shows clarity closer to and farther from a lens at sharpest focus. The zone increases as the lens aperture is reduced (often accomplished by increasing light level) or the lens angle is widened.

Ediophor A commercial-grade large-screen video projector made in Switzerland that utilizes an RGB color system to produce a video image electronically by etching the images into an oil film that modulates a high-intensity xenon light source.

End-stage The placement of a portable stage in an open, flat-floor building centered on the short end to give maximum seating capacity.

Equipment manager Acceptable professional term for the person who sets up and maintains the band equipment.

Exposure The selective control of reproduced tonal values within a system's limits. *Overexposure* results when reflected light exceeds the camera's limits. *Underexposure* results when a surface is insufficiently bright to be clearly discerned in the picture.

Fill light The supplementary illumination used to reduce shadow or contrast range.

Flash button On a lighting console, a momentary contact switch that allows a control channel to bypass the fader and instantly bring the circuit to 100 percent output. Normally found on consoles designed for rock and roll applications.

Followspot A high-power, narrow beam light suited for long throws (typically 100 to 300 feet), generally with an iris, shutters, and color changer. It is designed for hand operation to follow the movements of performers.

Found space Any space used for theatre, concerts, or television productions not specifically designed as a performance area.

f-stop A measure of the light-transmissive ability of a camera or projection lens.

Gobo A pattern or breakup placed in front of a hard-edged light designed to cast a specific design or modeling onto a surface.

Grand tour A tour taken by a star, usually with only a piano, to perform excerpts of famous opera and classical works.

Groupies People, often girls, who are so devoted to a popular recording artist that they collect memorabilia, join fan clubs, and go out of their way to be everywhere the artist is.

Hanging point The point on the truss from which the attachments are made for rigging.

High density dimming The use of advanced engineering and microprocessors to miniaturize the electronic dimmer.

HMI Metal halide lamp (mercury/argon additives) in tubular, double-ended form at 5600 Kelvins.

Hod A number of individually jacketed cables, usually three #12 wires, taped together for easy transport and layout on a lighting pipe.

HTI Metal halide short-arc lamp at 5600 Kelvins.

IATSE The International Alliance of Theatrical Stage Employees. The international bargaining unit for stagehands such as property masters, audio technicians, carpenters, riggers, electricians, wardrobe mistresses, and other backstage employees.

Iso An abbreviation of the word *isolated*. In television it denotes the switching of a camera onto a tape reel other than the master tape.

Kelvin temperature The unit of temperature used to designate the color temperature of a light source.

Key light A motivating source of illumination that establishes the character and mood of the picture.

Key word A method of calling cues that does not rely on visual or mechanical signaling devices. A word such as "go" or "out" prompts someone to react in a planned manner.

Lag A persistent or smearing afterimage on a television screen following the image of a moving object.

Layering The use of color to create depth and separation by using different shades or saturations of a single color.

Letter of agreement A legal agreement generally written in plain language based on a business letter form rather than a legal form.

Light show A mixture of theatrical lighting, projections, film, and black light images used to create an environment for the audience (popular in the 1960s).

Looks Planned patterns of light, and often color combinations, that are programmed and used one or more times in a concert.

Lumen per watt The number of lumens produced by a light source for each watt of electrical power supplied to the filament.

Lumens A unit of (light) flux.

Luminance The true measured brightness of a surface.

Matrix or pin matrix A device used to group channels to one or more master controllers, often employing a small pin or patch placed on the console.

MOR *Middle-of-the-road*, designating music that is marketed between easy listening and rock and roll, such as Barry Manilow, Air Supply, or John Denver.

Multicable More than one complete electrical circuit housed in the same flexible protective jacketing.

Off-line editing Preliminary editing of videotapes, using copies of the original tapes, usually performed on a low-cost ¾-inch editing system that allows the director to make editing experimentations before the final decisions without expensive edit bay and editor costs (often called a *work tape*).

On-line editing Final editing of videotapes, using the original master tapes to produce a finished program.

Package dimming The grouping of typically six or twelve dimmers in a single integrated enclosure, often used for portable work.

PAR-64 A *Parabolic Aluminum Reflector* light used with a quartz lamp sealed behind one of a group of lens configurations. The *64* designates the diameter of the lens in eighths of an inch.

Patching The act of interconnection of dimmer output to loads via a patch panel, a system that usually uses a plug and jack on the line side of the load only.

Per diem A daily allowance paid by an employer to the crew and cast while they are away from home, used to cover nonreimbursed expenses such as food and laundry.

Pickup point The point of attachment to the building's load-bearing structure used for rigging.

Pigtails Short 3 ft to 5 ft cables, usually #2/0 or #4/0 welding cable, used to connect to the house power panel.

Pink contract An agreement issued by the IATSE to qualified members who want to travel outside of their local jurisdiction.

Pin matrix See Matrix.

Prep The time devoted to organizing and packaging materials such as lighting or sound prior to the rehearsal or tour, usually in a location other than the first venue.

Programming The real-time mixing of multiple video camera shots or taped feeds onto one master tape or into a live line transmission.

Proscenium An arch placed so as to separate the audience from the acting or performance space.

Psychovisual Relating to the psychological use of color to effect emotional response in the viewer.

Punch light A lighting fixture with a highly concentrated beam allowing long throws.

Ray light A reflector with a separate 120 volt lamp, usually of 600 watts, which fits into a PAR-64 housing. This unit creates a very narrow beam of light, the same as an ACL but without the 12 volt problem of matching to 120 volt dimming.

Rigging The process of installing lines or motors to support trusses, scenery, et cetera, in their needed positions.

RMS *Root Mean Square*. A method of averaging a sine wave to arrive at a mean average.

Road company The authorized production of a play or musical that is produced in one city and is then performed in multiple cities for limited engagements around the country.

Roadies An early term used for the people who traveled with popular bands taking care of the musical equipment.

Saturation The purity of a color; the extent to which a color has been diluted.

SCR (silicon-controlled rectifier) dimmer A solid-state electronic device used to control lamp brightness by cutting off part of the cycle or the alternating current supply in a specific type of dimmer.

Scrim A gauze-like curtain that when illuminated from the front appears opaque, but if light is brought up behind the scrim becomes transparent.

Set The order of songs to be performed, generally used for musical performances where no dialogue or scripted words are recited.

Side light The source used to rim faces and model profile shots, often at or near head height.

Soft patch An electronic means of assigning dimmers to control channels without physically wiring jumpers on a terminal board or patch panel.

Span set A circle made of nylon web strands encased in a nylon covering which is wrapped around areas of load-bearing structures to prevent chafing.

Streaking The well-defined light or dark horizontal stripes super-imposed across a picture, resulting from an area of extremely high tonal contrast in the picture.

Theatrical smoke Smoke created either through a chemical reaction or combustion (often caused by heating an oil-based liquid until it vaporizes).

Torm A slang word for "tormentors," which are the side curtains that hang next to the proscenium opening. These curtains can be adjusted to alter the width of the performance area.

Trade usage A custom or widely used practice common to a business, unwritten but believed to be generally understood.

Truss A metal structure which is designed to support a horizontal load over an extended span. In theatre and concert work, it is the term applied to the structure that normally supports the lighting over a 40 to 60 foot clear span.

Tungsten-halogen lamp (Quartz) A type of lamp design achieving an almost constant output and color temperature with a higher lumen output through the working life of the lamp. This results from the halogen vapor (or bromine or iodine) that facilitates a chemical recycling action, preventing the blackening of the bulb wall with filament particles.

Turnaround In music, a point in the tune when the melody is broken up by another musical idea, after which the original melody is repeated.

United Scenic Artists (USA) A national trade union that represents theatrical designers such as set designers, lighting directors, scenic artists, and model makers.

Venue The location of a gathering; often used in legal contracts to designate the location of the concert.

Violent failure The term most often used to describe a lamp exploding.

Wire rope Similar to aircraft cable but generally with less tensile strength.

Xenon A compact source discharge lamp containing xenon gas in a pressurized lamp housing.

Yellow card A term used by the IATSE to designate shows in which all the technicians are members of the union working under a signed agreement for wages and working conditions, sanctioned by the International office.

Index